Thirst

Filippo Menga is an Associate Professor of Geography at the University of Bergamo and a Visiting Research Fellow at the University of Reading. Editor-in-Chief of *Political Geography* and recognized internationally as one of the leading experts in water politics, he is the author of *Power and Water in Central Asia* and co-editor of *Political Geography in Practice and Water, Technology and the Nation-State*. He serves on the editorial boards of *Central Asian Survey*, *Global Networks* and *Studies in Ethnicity and Nationalism*.

Thirst

The Global Quest to Solve the Water Crisis

Filippo Menga

VERSO

London • New York

First published by Verso 2025
© Filippo Menga 2025

All rights reserved

The manufacturer's authorized representative in the EU for product safety (GPSR) is LOGOS EUROPE, 9 rue Nicolas Poussin, 17000, La Rochelle, France
Contact@logoseurope.eu

The moral rights of the author have been asserted

1 3 5 7 9 10 8 6 4 2

Verso
UK: 6 Meard Street, London W1F 0EG
US: 207 East 32nd Street, New York, NY 10016
versobooks.com

Verso is the imprint of New Left Books

ISBN-13: 978-1-80429-071-2
ISBN-13: 978-1-80429-073-6 (US EBK)
ISBN-13: 978-1-80428-072-9 (UK EBK)

British Library Cataloguing in Publication Data
A catalogue record for this book is available from the British Library

Library of Congress Cataloging-in-Publication Data
A catalog record for this book is available from the Library of Congress

Typeset in Sabon by MJ & N Gavan, Truro, Cornwall
Printed and bound by CPI Group (UK) Ltd, Croydon CR0 4YY

Emilia Sofia, questo è per te

Contents

Preface ix

Introduction: Crisis 1
1. Crisis Management 27
2. Care: Good Samaritans and the Politics of Care 59
3. Sacrifice: #WaterCrisis – Utopias and Relational Sacrifice 95
4. Corporate Social Redemption: Common Water, Bottled Profit 125
5. There Is No Such Thing as a Global Water Crisis 161

Further Reading 177
Acknowledgements 181
Index 183

Preface

There has never been a time in human history when the water crisis has been so central to political, cultural and economic debates. But how do we talk about it? And who are we listening to? In other words, whose voices are the loudest? The answer to this last question can be surprising. Take the recent World Water Forum (WoWF) – the largest international event about water – held in Bali, Indonesia, in May 2024. You might think that politicians, or perhaps scientists, would have dominated the stage. That wasn't the case. Instead, the main event, the one that really grabbed the headlines and will be remembered in the future, was the keynote speech given by Elon Musk, one of the richest people in the world, who has no particular expertise in water-related issues. According to Musk, the water crisis is solvable. His solution? Solar-powered desalination plants. I told my barber about this, and he agreed with Musk. And other people I have recently encountered, including someone at the garage, my tennis partner and a twelve-year-old boy, also agree with Musk's vision. Is Musk right? He is clearly wrong. Desalination is extremely energy-intensive (and even if – and I must underline *if* – a plant manages to run only on solar power, that means there is less energy available overall for all the other things that humans need energy for), it makes water pricier, and scientists have yet to find a way to deal with the highly toxic seawater desalination brine resulting from the process. Anyone can have an informed opinion, but if you are 'a rich guy with an opinion', as Bill Gates has called himself, people do listen to you. Yet the loudest voice isn't always the right one.

This is not a book about Elon Musk or about any specific individual, even though several other well-known figures who have taken an interest in the water crisis will be discussed in detail. This is a book about water. And in particular about how capitalism has noticed water.

I write as people in the northern hemisphere prepare for a summer of water restrictions. Many artificial reservoirs and underground aquifers in Europe and the United States are below warning levels. Ruins that were once buried by lakes are reappearing. What is happening in parts of the world in 2024 is reminiscent of what happened in Cape Town in early 2018, as the South African city awaited the dreaded Day Zero, the day when municipal water was expected to run out. At the time, I was teaching a course on water politics at the University of Reading, and I made a habit of commenting on the situation in South Africa before my weekly lecture. My – mostly British – students had never experienced a drought and found it astonishing that people in Cape Town had to queue for hours to fill a couple of 10-litre water cans and, more generally, that a city could 'run out of water' at all. Growing up in Sardinia in the 1980s and 1990s, I was not surprised by the idea of water rationing. One of my students explained that his brother was in Cape Town on business at the time. Apparently, for him and others like him who were staying in one of the city's luxury hotels, water was not restricted. The showers were working and even the swimming pool was available. Obviously, for some, Day Zero had already arrived, and for others it might never come. What did come, in June 2018, was the rain. Danger was averted for Cape Town.

Ironically (for me and my students, at least), the summer of 2018 was also the summer when the drought arrived in the UK. In the south of England, where I lived, there was no rain for nine consecutive weeks, between May and July, a phenomenon unprecedented in recent times. I vividly remember the smell of the sun-baked earth and the yellowish colours of the landscape, which, week after week, reminded me more and more of my beloved Sardinia. In the park near my home, neighbourhood committees organized shifts to take a few buckets of water to the thirstiest trees, as the government once again proved unprepared to tackle one of its many water crises. English lawns, renowned for their ability to stay green without the need for irrigation systems, turned into fields of scrub, as did most of Reading's many parks. Water was not rationed, but it came close. Then, messianically, the rain came.

Today, water is not something you can take for granted. This is well known to the inhabitants of Mexico, who since 2023 have been facing a drought that has gone from being exceptional to semi-normal (but what is normal these days?), or to those of Emilia-Romagna (Italy) and Lower Saxony (Germany), areas devastated by severe floods in 2023 and 2024, respectively. The climate crisis is also a water crisis. Global warming is accompanied by extreme weather events such as droughts and floods; it is more difficult to predict rainfall or recognize the seasons; and melting glaciers are changing the seasonal regimes of the world's waterways. And this, according to many observers and scientists, is the meaning of the term 'Anthropocene', an unstable geological epoch in which humans are coming to terms with the consequences of their actions. The environmental crisis, and with it the water crisis, is no longer something that may happen in the distant and indefinite future. The crisis is now, and it is unpredictable. And precisely because the water crisis affects a large part of humanity, we can go so far as to say – with the necessary clarifications that I will make later – that we are facing a global crisis.

Now, you may ask, what are we doing, or rather what can we do, to solve this problem? According to what large transnational companies and many newspapers tell us, most of the responsibility for action lies with citizens: 'take shorter showers'; 'use the eco programme on your dishwasher'; 'turn off the tap when you brush your teeth'; 'buy sustainable'; and 'donate to a charity that will do it for you'. These are some of the messages that we are exposed to on a daily basis. But is it possible to solve a structural problem with individual actions? In other words, is it possible to cure the planet by working on the symptoms rather than the causes of its malaise? According to Jean Baudrillard – one of the most original and prophetic intellectuals of the twentieth century – the hallmark of contemporary society is that 'consumption of consumption' is more important than consumption itself. We live in a society that consumes itself and, in the process, consumes the planet.

When – years ago – I decided to write a book about the water crisis, I did not think that I would be writing a book about capitalism and its illusions (or deceptions). Today, I think that you cannot write about the water crisis without talking about capitalism. Writing about capitalism means writing about what is undoubtedly the most important economic, political, social and cultural force of our time. Not to do so might have been easier and certainly quicker. But that would also have implied dealing with only part of the problem, only the obvious and visible. So, since writing a book takes time and effort, I thought I might as well tell a full story. Or at least try.

Introduction: Crisis

In 2018 something caught my attention: a water charity, Water.org, partnered with Stella Artois, the iconic beer brewer, to produce a Super Bowl commercial called 'Taps'. In the commercial, Hollywood superstar Matt Damon told viewers that they should not take water for granted. Tap water is a privilege, and millions of people around the world do not have access to clean, safe or reliable water sources. But a beer glass sold by Stella Artois could change that, said Damon: 'If just one per cent of you watching this buys one, we could give clean water to one million people for five years.' While I agreed with the underlying message about the importance of basic water access, a few things puzzled me. Why was Damon talking about water, and where did he get those figures? How can a glass help us solve the water crisis? And why is this message being voiced during a commercialized spectacle? As often happens, those questions raised a few more, and my modest attempt to answer them produced this book.

Charity and the idea of giving are seemingly distant from the capitalist logics of growth, accumulation and profit-making, but these logics pervade our lives. Capitalism encompasses and engulfs most processes in society, and this also shapes the ways in which we are confronting the so-called 'global water crisis'. And so, if the Super Bowl is one of the biggest, if not the biggest, spectacles of capitalism, we should not be surprised that the water crisis has become part of the show – even more so if we consider that the water crisis is arguably a product of capitalism.

Crisis, in turn, is a key feature of capitalism. It is arguably its structural foundation. Cyclical and periodic crises such as the 2007–8 financial recession are not one-off occurrences but rather the result of one of the underlying contradictions of capitalism: the accumulation of capital and its appropriation of surplus value happens at the expense of workers, whose diminished means of consumption leads to a contraction in demand. While production grows, consumption cannot follow at the same pace. The environmental crisis is also a result – and a further complication – of this process, through the expansion of what Jason Moore called the commodity frontier: capital must expand and commodify nature to perpetuate itself and satisfy its metabolic demand for cheap food, labour, energy and raw materials. It is now a structural occurrence, rather than an exceptional event, and profit-making management of the environmental crisis has become an integral part of the workings of capitalism. In other words, as it becomes clear that capitalism does not work for our ecosystems, the ecosystems – and water itself, as this book will show – are supposed to create profit and, therefore, reproduce capitalism.

If someone opens a newspaper, they are likely to find stories about the water crisis. We are apparently destined to fight wars over 'blue gold', and journalists are eager to report how this story unfolds: the water crisis in Flint, Michigan, and its linkages with systemic racism in the United States; the deterioration of water quality due to widespread water pollution; the impending deluge of water from melting Central Asian glaciers and in several other areas around the world; the 'Day Zero' narrative used to dramatically anticipate the end of urban water supplies in Cape Town and elsewhere since 2018; billions of people who lack access to water and sanitation, and billions that are already experiencing – or

INTRODUCTION

will experience in the near future – absolute water scarcity.¹ I could carry on. The sense of urgency about the water crisis has entered popular and political discourse and is widespread. In humanity's quest to solve the 'global water crisis', we are all called to rise to the challenge.

The 'global water crisis' thus flows into our lives: our relationship with it is intimate, built along a thin line that unites – but also separates – corporate, individual and collective responsibility. I could certainly do more to save water and, in case I forget, I just need to turn on the TV. Super Bowl commercials provide, again, another fitting example. In 2016, the toothpaste giant Colgate purchased a thirty-second ad to encourage us to turn off the tap while we brush our teeth. In it, we see a man standing in a bathroom: 'Brushing with the faucet running', the ad says, 'wastes up to 4 gallons of water. That's more than many people around the world have in a week. Please turn off the faucet. Spread the word. #EveryDropCounts.'

It is difficult to disagree with this message. Yes, we should turn off the faucet while brushing our teeth, or for that matter, take shorter and less frequent showers, and be more conscious as consumers. What is notable here is the social angle of the campaign. Colgate asks consumers to 'share how your own small actions are making a difference by using #EveryDropCounts in your social posts'. The mini-website everydrop-counts.org reports that we can save up to sixty-four cups of water every time we brush our teeth. Using data sourced via a consumer survey, Colgate calculates that 'people's actions in response to Colgate's Save Water

---

1 A community faces absolute water scarcity when their annual water supplies are below 500 m³ per capita per year. To contextualize this, water withdrawals per capita in the United States (in 2022) amounted to 1,207 m³, in India (in 2021) 541 m³, in Australia (in 2022) 631 m³ and in China (in 2022) 398 m³.

campaign have led to the potential reduction of 99 billion gallons of water'. But correlation does not imply causation. If, on one hand, the ad noted that many people around the world have very limited access to water, then, on the other hand, it is unclear how those sixty-four cups of water that someone saved, say, in the United States, are helping those people or are improving water access in countries which lack basic water services, such as Eritrea or Papua New Guinea. Other than raising awareness of an important matter and asking consumers to promote the company on social networks for free, it seems that saving water is not actually what this ad is about.

In a similar vein, in 2020 the Italian edition of *National Geographic* launched a website for a campaign aimed at raising awareness of the water crisis in Italy and giving recommendations on what can be done to solve it. The website explained that climate change is having an impact on the availability of freshwater resources in Italy, and future generations will pay the price for this unless we act immediately. The website features a few videos and articles that showcase innovative ways to save water and promotes six practices that we can adopt to 'make a difference'. One of these is using a dishwasher, as this is more efficient than washing dishes by hand. This *National Geographic* campaign is sponsored by Finish, a company that sells a range of dishwasher detergent and cleaning products. The *National Geographic* website also links to another website created by Finish (acquanellenostremani.it), 'water in our hands'. Visitors are prompted to take a test to find out how each of us can live more sustainably and make a 'pledge for water'. I took the test, and my water consumption amounts to roughly 316 litres per day. After this, the webpage asked me to check three boxes, promising: to take shorter showers; not to rinse my dishes before putting them in the dishwasher; and to run my washing machine only if fully loaded. This then led me to promise to save water and 'improve my habits', offering me the opportunity to share my pledge on Twitter and Facebook. I declined, feeling a little guilty about my water consumption.

Clearly, as the author of yet another book on water, I am inclined to pay attention to all water issues. And yet, the water crisis is no longer – and not only – a problem for water experts.

We are led to believe that the 'global water crisis' unfolds in our kitchens and bathrooms, and it has to do with our habits and small excesses, such as taking a longer shower every now and then. As a consumer, I could always do better. But can a structural and systemic problem be solved by individual action? And is the act of sharing a hashtag on social networks a driver of change, or rather a self-referential display of environmental virtuosity? And should I really be thinking of myself as a consumer, or, at least, only as a consumer? The range of options that this leaves me with is quite limited and is certainly not going to change the fact that 42 per cent of the water that enters the Italian network is lost due to leaks and therefore cannot reach households (and my dishwasher), or that 29 per cent of the Italian population does not trust the quality of tap water.[2] And will taking shorter showers challenge the reality of Italy as Europe's biggest consumer of bottled water and second in the world, behind only Mexico, with one of the biggest bottled water businesses in the world?[3] Clearly, the concept of having a 'water footprint' needs to be carefully interrogated, and so does shifting the responsibility for the water crisis and ecological catastrophe to the individual. As geographer

---

2 ISTAT, *Utilizzo e qualità della risorsa idrica in Italia*, Rome, 2019.

3 UCIMA (Unione Costruttori Italiani Macchine Automatiche. per il confezionamento e l'imballaggio) *Surge in Italian Bottled Mineral Water Consumption*, press release, 12 April 2018.

Matthew Huber has pointed out, the idea of an 'ecological footprint' puts the blame for ecological degradation on consumers. This approach revolves around the message 'that we need to live simply and consume less', but this does not really challenge the source of the crisis: capital.[4]

Yes, our conscious consumption can have transformative effects across the world. However, the problem with this logic is that not everyone can afford to consume consciously, and that the hypothetical redistributive effects of consumption from rich to poor countries are hard to measure or verify. Regardless of this, our planet is sick. And thirsty. Everyone wants to play a role in saving the planet and solving the water crisis, and the corporate world is eager to join in. This is the story I want to tell in this book.

Why another book on water? We are surrounded by crises, catastrophic events and premonitions of tragedy. In 2020, with the global Covid-19 pandemic, some of our worst nightmares came true, and yet, only a few years later, the pandemic seems to have been partly forgotten. Many people, particularly in rich countries, protested against lockdowns, mandatory face masks and Covid-19 vaccines – all measures aimed at protecting the most vulnerable – which were perceived as a violation of individual freedoms. In 2022 Russia invaded Ukraine, bringing war to the doorstep of Europe with a virulence that to many resembled the harbingers of World War II. As Russian gas disappeared from their energy mix, European countries braced for an energy crisis. In April 2022, Italy's Prime Minister Mario Draghi asked its citizens: 'Do you want air conditioning or peace?' Of course, people chose air conditioning, and in winter the heating was on as usual. Unsurprisingly, energy prices went up, and this was an issue for the poorest, but life went on following its

---

4 M. Huber, 'Ecological Politics for the Working Class', *Catalyst* 3: 1 (2019).

usual routine. We might have supposed that renouncing Russian natural gas, oil and coal would have accelerated the EU's transition towards renewable energy; instead, it only led to the strengthening of alternative gas import routes – primarily through North Africa – and the reopening of coal-fired power plants in the supposed European environmental champion, Germany. In March 2023, also following what has become a familiar routine, the Intergovernmental Panel on Climate Change (IPCC) delivered its recurring final notice: act now or it's too late! And even though the UN secretary-general, António Guterres, added that 'our world needs climate action on all fronts: everything, everywhere, all at once', a privileged part of humanity prefers to manage its retreat (take, for instance, the purposeful and coordinated movement of people and buildings away from risk, as is being done in costal Florida) rather than address the root causes of climate change, while those who are not afforded this option will be left behind.

The climate crisis is also a water crisis. Global warming is accompanied by extreme weather events such as droughts and floods, and it is harder to predict rainfall and seasons. As a consequence, it is more difficult to store water and practise both irrigated and rainfed agriculture. Water crises are thus everywhere. But while we have been talking about water problems for decades already, we can safely say that it is with the two severe droughts that affected Europe and North America in both 2022 and 2023 that 'water fever' picked up. This is when many of the news outlets in the Western world started to relentlessly cover the water crisis, triggering a chain reaction that resulted in the proliferation of high-level conferences on water, countless round-tables and workshops, thousands of policy papers and policy briefs on water, and an uncountable number of declarations and opinions about what the water crisis is and how to solve it. The fact that everyone – even countries in the Global North that had so far been sheltered from it – is now exposed to the water crisis and the climate crisis shifts anxiety from the domain of the extraordinary to that of the ordinary and manageable. Opinions on carbon capture strategies, sea level rise, droughts, or desalination can also be heard (and voiced) in parliaments and general assemblies, but also in gyms, bars and restaurants.

In the fifteen years or so that I have been working on water issues, I have become familiar with the various discourses on water that have been mobilized. As I mentioned earlier, journalists have an interest in 'water wars', and more generally in anything related to conflicts and specific crises (exceptional droughts, floods, pollution scandals and so on). Politicians are increasingly worried about transboundary water disputes and water security – worried, that is, about the impact of water scarcity on health and livelihoods and the potential social and political unrest that this might trigger. Private companies are concerned about water primarily when the resource can be commodified or when a shortfall in its supply might lead to an economic loss. All roughly agree on the fact that the solution to water problems is: more resilient infrastructure (more – and interconnected – reservoirs, better water treatment plants and less leaky pipes); technological solutions (more efficient irrigated agriculture, smart water meters, less water-thirsty GMO crops, and desalination plants); stronger and more equitable legal instruments (particularly when it comes to transboundary water disputes); smaller human populations (although this is more contentious[5]); and the ever-popular,

---

5 Population is often mistakenly blamed as the main driver of biodiversity loss and environmental degradation, while this is generally due to broader consumption patterns. I did write about this, together with several biologists, in the journal *Biological Conservation*: A. C. Hughes et al., 'Smaller Human Populations Are Neither a Necessary nor Sufficient Condition for Biodiversity Conservation', *Biological Conservation* 277 (2023).

and yet very vague, 'better governance'. Academics, in turn, provide a much more nuanced picture. In addition to specific water-related disciplines such as hydrology and hydraulic engineering – and indeed technical studies on water management and water treatment abound – many other disciplines in the social sciences and humanities have taken an interest in water, looking at the political, economic, cultural, legal and spiritual value. Indeed, water provides an excellent lens through which some of the contradictory and often unequal dynamics that shape social interactions can be interpreted and explained. And the relationship between humans and water is not static. It is a process and, as with any other process, it is constantly evolving and changing.

However, quite astonishingly, there is one thing that has not entered the debate, and it is the fact that talking about the water crisis is not very different from talking about the current status of capitalism. Thus, in spite of the wealth of books about water released each year, few, if any, have looked into how the water crisis has changed and evolved, and how it has become a common, and highly contradictory, popular concern that speaks to the absurdities of late capitalism.[6]

Two other events were unfolding around the time the Water .org Super Bowl ad aired: the South African city of Cape Town was quickly running out of water, and people were

---

6 If we look at 2023 only, for example, two books deserve to be mentioned. The first is Mark Zeitoun's *Reflections*, which looks at the mutually constitutive relationship between water and societies (*Reflections: Understanding Our Use and Abuse of Water*, New York: Oxford University Press, 2023). The second is *Water: A Critical Introduction* by Katie Meehan, Naho Mirumachi, Alex Loftus and Majed Akhter, Hoboken: Wiley & Sons, 2023, a textbook that examines water through the hydrosocial cycle, the idea that water is inseparable from society, and that 'water shapes – and is shaped by – social practices and geometries of power'.

taking to the streets in the Mexican border town of Mexicali over a deal between the Mexican state of Baja California and Constellation Brands, the third-biggest brewer in the US. While the two events are not directly connected, they both speak to some of the challenges and contradictions that I have outlined.

The Cape Town drought began in 2015, but escalated into a severe water crisis in late 2017 and peaked in June 2018, when strong rains began filling the almost empty reservoirs serving the city, thus avoiding the arrival of the much-dreaded Day Zero. It was the City of Cape Town itself that proposed the apocalyptic idea of Day Zero, making it the first major city in the world at risk of running out of water. Even though the day was initially set for 22 April 2018, the implementation of restrictions on water consumption slowed down reservoir depletion, enabling the City to push Day Zero back to 28 June 2018. Reservoirs can indeed ensure water availability while also increasing a community's dependence on said infrastructure, a paradox that Giuliano Di Baldassarre refers to as the 'reservoir effect'.[7] In early 2018, near the peak of the crisis, the City of Cape Town constructed emergency desalination plants at Strandfontein and Monwabisi to increase water supply. Desalination is an energy-intensive and costly process, and soon after the crisis was averted the city decommissioned these desalination plants, only to announce the construction of a large permanent plant in 2021 as a solution to future water crises. But these technical fixes can easily backfire. Giorgos Kallis explains that an increase in water supply generates higher demand, which in turn favours supply expansion over other alternatives, triggering a vicious cycle that expands the water footprint of cities.[8] Furthermore, and beyond the supply and

---

7 G. Di Baldassarre et al., 'Water Shortages Worsened by Reservoir Effects', *Nature Sustainability* 1: 11 (2018), pp. 617–22.

8 G. Kallis, 'Coevolution in Water Resource Development: The

demand cycle, an increase in water availability can lead to substantive changes in behaviour. With the advent of large-scale desalination in Israel, water consumption went up even if prices also increased.⁹ In 2018, when Israel was facing the fifth year of a devastating drought, the Israeli Water Authority launched a public campaign to remind residents that they still had to save water, despite the country's large desalination capacity.

Day Zero soon became a major global news headline, and journalists and politicians widely mobilized this narrative to raise a stark warning of what awaited the rest of the world unless we acted quickly and responsibly. As two researchers of the crisis put it, 'Day Zero clearly showed just how serious the situation was, and located Cape Town's experience of scarcity in a broader planetary geography of climate change and impending climate crisis.'¹⁰ But the Cape Town water crisis was not only due to a drop in precipitation. Poor planning, administrative disconnects and high flows of migrants from Southern Africa all contributed to exacerbating the crisis. Above and beyond all this, however, Elisa Savelli and her co-authors have shown that it was the unequal distribution and unsustainable consumption of water in upper- and middle-class suburbs that worsened the negative effects of the drought across the city.¹¹ At the peak of the crisis, the

---

Vicious Cycle of Water Supply and Demand in Athens, Greece', *Ecological Economics* 69: 4 (2010), pp. 796–809.

  9 D. Katz, 'Undermining Demand Management with Supply Management: Moral Hazard in Israeli Water Policies', *Water* 8: 159 (2016).

  10 N. Millington and S. Scheba, 'Day Zero and The Infrastructures of Climate Change: Water Governance, Inequality, and Infrastructural Politics in Cape Town's Water Crisis', *International Journal of Urban and Regional Research* 45: 1 (2020), pp. 116–32.

  11 E. Savelli et al., 'Don't Blame the Rain: Social Power and the 2015–2017 Drought in Cape Town', *Journal of Hydrology* 594 (2021). See also E. Savelli et al., 'Urban Water Crises Driven by Elites' Unsustainable Consumption', *Nature Sustainability* 6 (2023), pp. 929–40.

mostly white city elite had to reduce its water consumption, but nevertheless was still able to use considerably larger amounts of water than township dwellers and people in the lower and upper middle classes. Despite the strong legacy of colonialism and racial segregation, and an uneven water distribution system, the Day Zero narrative 'shifted the burden of accountability from the city to the citizens, who suddenly became responsible for drastically reducing their consumption and avoid Day Zero'.[12] Again, we are walking along that thin line that separates individual from collective responsibility. What shook the world about the Cape Town water crisis was not necessarily the fact that the city was running out of water. Many people in Cape Town had been suffering from chronic water insecurity for decades. Rather, it was the fact that the drought affected a large and wealthy part of its population for the first time in history.

Ten thousand miles away from Cape Town, another water crisis was unfolding in Mexicali. In this case, the reason was not a drought, or at least not only a drought. Three years earlier, in 2015, Constellation Brands – a large beverage alcohol company based in the United States – struck a deal with the governor of the Mexican state of Baja California to construct a large brewery in Mexicali, on the border between southern California and Mexico. While the deal had been negotiated in secrecy, in its press release the company remarked that Mexicali 'is ideally located near the US state of California, Constellation's largest beer market. Initially, the brewery will be built to provide 10 million hectolitres of production capacity with the ability to scale to 20 million hectolitres in the future.'[13] This location is 'ideal' to

---

12 Ibid., p. 6.
13 Constellation Brands, *Constellation Brands to Build New 10 Million Hectoliter Brewery in Mexicali, Mexico and Further Expand Its Nava Brewery to Fuel the Continued Industry-Leading Growth of Its Beer Business*, press release, 7 January 2016.

Constellation as it would award the company rights to draw from the Mexicali Valley aquifer, rather than from other aquifers upstream of the Mexico–United States border. The environmental costs of the brewery would thus be borne by Mexico, while the economic gains would be transferred to the United States.

The Mexicali Valley aquifer is recharged in large part by the waters of the Colorado River, or rather, it used to be. In his classic book *Cadillac Desert*, Marc Reisner tells the story of how, during the twentieth century, large hydraulic infrastructure such as the Hoover Dam and the Glen Canyon Dam, together with intensive agriculture and urbanization in California, Arizona and Nevada, made the Colorado River one of the most controlled (and litigated) rivers in the world.[14] The river is heavily overexploited and, as a result, the aquifer is also overdrawn and loses 456 million cubic metres of water per year, a deficit that would have been further exacerbated by the brewery.[15] In this drought-stricken region of Mexico, farmers and residents had been suffering from water shortages for years, and the planned brewery prompted a passionate backlash. Starting in 2017, and united around the movement Mexicali Resiste, thousands of people took part in demonstrations outside state government offices and the brewery construction site, blocking the construction of water pipes to the factory and venting decades, if not centuries, of frustrations provoked by unequal water abstraction between Mexico and the United States. To put things into context, it is worth remembering that in 1895 the attorney general of the United States, Judson Harmon, prepared an opinion on a territorial dispute between Mexico

---

14 M. Reisner, *Cadillac Desert: The American West and Its Disappearing Water*, New York: Penguin, 1993.

15 A. A. Cortez Lara, 'Elements of Socio-Environmental Conflict: The Constellation Brands Brewery and Mexicali Water', *Frontera norte* 32 (2020).

and the United States over the use of the waters of the Rio Grande. The opinion, which became known, notoriously, as the 'Harmon Doctrine', holds that a country is absolutely sovereign over the portion of an international watercourse that runs within its borders.[16] According to Harmon, the United States was entitled to full control and ownership of the Rio Grande, even if that meant that the river might disappear before reaching the Mexican border. Today, the Harmon Doctrine is considered anachronistic and irrelevant to international water law, and the government of the United States might superficially appear as more co-operative than it was in the past. But the plans and actions of a company like Constellation reiterate and give continuity to this history of exploitation and subjugation.

In a context in which the government of Baja California had already tried to pass a water privatization law in 2016, this further development was just too much. Protests intensified in 2017, and in a few instances policemen used their batons on demonstrators. The popular outcry over the brewery yielded its first results in 2017 when Francisco Vega, the governor of Baja California, abrogated the controversial water privatization law that had been passed behind closed doors and amid corruption rumours. In 2020, after a two-year boycott of Constellation, the state government called a referendum on the brewery, and 76 per cent of voters rejected its construction. Given the overwhelming result, in 2021 Constellation announced that it was giving up on its plans for a new brewery – at least for now.

As Jesus Galaz Duarte from the Mexicali Resiste movement explained, the brewery is a model of exploitation:

---

16 S. C. McCaffrey, 'The Harmon Doctrine One Hundred Years Later: Buried, not Praised', *Natural Resources Journal* 36 (1996), pp. 725–67.

When the market grows and has to satisfy consumers, they're going to deplete the water here. So what's going to happen? They're going to go to another place where there's more water to satisfy the same market and deplete their water. They're going to leave this region without the resources to live a dignified life.[17]

The story of the Constellation brewery is a good example of the contemporary socio-ecological upheavals that occur as a result of the expansion of the capitalist frontier. It is also testimony – if linked with the Water.org campaign in partnership with Stella Artois with which we began – of the sometimes absurd and often contradictory corporate and philanthropic responses to the 'global water crisis'. Humanity's pursuit of sustainability in one place, as Erik Swyngedouw puts it, 'is predicated upon the production of unsustainability elsewhere'.[18] The recent events in Cape Town and in Mexicali are interconnected, as they both show how the water crisis is more complex than the framing provided by corporate and philanthropic entities might lead us to think. Both events are a symptom of the unequal political and economic dynamics that shape water politics and determine the allocation and misallocation of freshwater resources. While the impacts of the water crisis tend to be unequally distributed and geographically localized – as they affect specific communities and social groupings – its root causes are structural. What is 'global' in the water crisis is the pervasive power of transnational corporations and the global political economy in which they operate. Consequently, the transformative effects of scattered technological fixes and decontextualized donations will stay within the boundaries set by the same

---

17 A. Zaragoza, 'As Big Beer Moves In, Activists in Mexicali Fight to Keep Their Water', NPR, 26 March 2018.

18 E. Swyngedouw, '"The Apocalypse Is Disappointing": Traversing the Ecological Fantasy', in H. Haarstad, J. Grandin and K. Kjærås (eds), *Haste: The Slow Politics of Climate Urgency*, London: UCL Press, 2022, p. 6.

structural conditions and exploitative relations that created the need for their very intervention. The interconnections of corporate and philanthropic responses to the 'global water crisis' unfold through a series of contradictions. This book is an attempt to recognize their shortcomings.

Humanity is, by all measures, in the midst of a water crisis. But while 'global water crisis' is a popular way to encapsulate our water concerns, this notion needs to be problematized and questioned. In an influential article published in 2001, David Demeritt shed light on how climate science is both epistemically and socially constructed, highlighting how climate change turned into a matter of global concern through the technocratic pressures of an international regulatory regime.[19] Likewise – and this is something that we will see in greater detail later in this book – we began talking about a 'global water crisis' in the late 1970s, even though water is not necessarily a 'global' resource, or at least, this globality is not that straightforward. Of course, water is geopolitical – consider, for instance, the politics of an international river like the Nile – and even universal but, at the supply level, water is an essentially local resource susceptible to monopolistic control. As Karen Bakker has observed, water is both transient and heavy, and thus expensive to transport relative to value, and 'uncooperative'.[20]

According to Jamie Linton, the social construction of the 'global water crisis' can be traced back to Soviet hydrology, which was driven by the need to produce hydrological information on the immense territory of the Soviet Union.[21]

---

19 D. Demeritt, 'The Construction of Global Warming and the Politics of Science', *Annals of the Association of American Geographers* 91: 2, 2001, pp. 307–37.

20 K. Bakker, *Privatizing Water: Governance Failure and the World's Urban Water Crisis*, Ithaca: Cornell University Press, 2010.

21 J. Linton, 'Global Hydrology and the Construction of a Water Crisis', *Great Lakes Geographer* 11: 2 (2004), pp. 1–13.

When the Soviet monolith collapsed in 1991, Soviet hydrologists began to apply their methods and models on a global scale, and questions related to water availability and scarcity gained traction as issues of global concern. These data and approaches, Linton continues, were then picked up in two extremely influential books – Sandra Postel's 1992 *Last Oasis: Facing Water Scarcity* and Peter Gleick's 1993 *Water in Crisis: A Guide to the World's Freshwater Resources* – and subsequently by the UN and other international organizations, which then led to widespread public attention on the idea of a 'global water crisis'. As Julie Trottier writes, while defining the 'water crisis' is a contentious matter, the idea that there is one which needs to be solved has imposed itself as a hegemonic concept, 'fuelling theories and policies that rely on its unquestioned existence'.[22] Where most analysts and policymakers agree, however, is on the fact that the world water crisis needs to be measured and quantified, as this is the precondition to fix it.[23] Such emphasis on measures and indicators empower the international 'expert': 'There is no room here to allow for a fundamental clash of values. The users are not expected to be active in proposing alternative values and their being responsible is measured as a function of their obedience to the experts' recommendations.'[24] As I have argued elsewhere, while the UN Sustainable Development Goals (SDGs) are an ambitious tool for collaboration aimed at improving the human condition (also with regard to 'SDG6: ensure access to water and sanitation for all'), their emphasis on quantitative methodologies and benchmarks does not capture the inherent politics and inequitable arrangements that often determine who does and who does not get access to water.[25]

---

22 J. Trottier, 'Water Crises: Political Construction or Physical Reality?', *Contemporary Politics* 14: 2 (2008), p. 197.
23 Ibid.
24 Ibid., p. 203.
25 H. Hussein, F. Menga and F. Greco, 'Monitoring Transboundary Water Cooperation in SDG 6.5.2: How a Critical Hydropolitics

Water is complex, and it is connected to pretty much everything that happens in the world. As such, it cannot be separated from social and political dynamics. Open any book about water and one of its first sentences is likely to remind you that water is essential for life and the development of societies, while also being irreplaceable and transient.[26] Indeed, there is no substitute for water, and the amount we have on our planet today is exactly the same amount that we had, for example, 5,000 years ago. We might be inclined to think (or at least, some of my students do) that water is renewable because of rainfall or seasonal water melt from glaciers, but water is a finite resource. And even though we can – reluctantly – try to find substitutes for other finite resources such as coal or oil, we cannot do so with water.

While it is difficult to verify such predictions, a recent study published in *Nature Sustainability* found that there is enough water to sustain life for over 10 billion people, and that, therefore, there should be plenty for the current population of less than 8 billion.[27] As a matter of fact, and in spite of the resurgence of neo-Malthusian theories that emphasize the limits of nature,[28] people around the world were already suffering from water stress or water-related diseases in the 1960s,[29] when the world population was just

---

Approach Can Spot Inequitable Outcomes', *Sustainability* 10: 10 (2018).

26 And indeed, I have replicated this scheme also in my first book: F. Menga, *Power and Water in Central Asia*, London: Routledge, 2018, p. 3.

27 D. Gerten, V. Heck and J. Jägermeyr, 'Feeding Ten Billion People Is Possible within Four Terrestrial Planetary Boundaries', *Nature Sustainability* 3 (2020), pp. 200–8.

28 In 1972, the publication of *The Limits to Growth* – a report commissioned by the Club of Rome – provided impetus to neo-Malthusian theories predicting the catastrophic consequences of resource scarcity.

29 This is when AQUASTAT, the FAO's Global Information

over 3 billion, and since 1900 more than 2 billion have been affected by drought. As Boretti and Rosa illustrate, from the 1950s global water withdrawals increased by roughly twice as much as the world population.[30] So, population is not the issue, or at least not an issue per se. It is how and what humans consume that matters. Yes, we must turn off the tap while we brush our teeth and we could take shorter showers, but overall, households and municipal use account for about 10 per cent of water use worldwide. If we really want to make a difference (at least on overall water withdrawals), we must focus on industrial water consumption (20 per cent of the total), and even more so on irrigated agriculture, which is the largest user of water globally and accounts for 70 per cent of water use worldwide.[31] Inefficient irrigation practices and intensive groundwater pumping deplete aquifers, while fertilizer run-off and pesticides can contribute to water pollution. Importantly, not all crops have the same social and environmental impact. For instance, the Western-led avocado boom – caused by rich countries' passion for wellness and superfoods – is depleting water resources in Chile, where the availability of cheap land and water led investors to convert potatoes, beans, corn and other crops to water-intensive and profitable avocado production in a

---

System on Water and Agriculture, started to collect data on water resources and agricultural water management.

30 A. Boretti and L. Rosa, 'Reassessing the Projections of the World Water Development Report', *NPI Clean Water* 2: 15 (2019).

31 This distribution (10, 20 and 70 per cent to, respectively, municipalities, industry and agriculture) varies across regions and continents, and can tell us a few things about their political economies. For instance, 44 per cent of total water abstraction in Europe goes to agriculture, 40 per cent to industry and energy production and 15 per cent to public water supply. Things are quite different in South Asia, where agriculture accounts for almost 95 per cent of total water use. See 'AQUASTAT – FAO's Global Information System on Water and Agriculture', available at fao.org.

water-scarce country.[32] Similar examples of unsustainable agricultural practices can be found in cotton farming in arid Uzbekistan and Turkmenistan and kiwi production in Italy, and the list goes on.[33] Demands for water vary across time and space, and while capital-driven intensive agriculture may indulge the market demands of a globalized economy, this often happens at the expense of small farmers, who no longer have enough water to cultivate their land or sustain their households, and thus have to leave and try to start a new life elsewhere, often in an urban area.

Furthermore, our freshwater resources are unevenly distributed. At the geographical level, certain countries have more water than others, and within countries, certain regions have more water than others, as in the case of the dry Northeast region of water-rich Brazil. At the socioeconomic level, income, class, race, gender and age might determine who has or does not have access to clean and safe water at a particular moment in time. More generally, the reports and data published by UN-Water – an interagency mechanism established in 2003 to coordinate the efforts of over thirty UN organizations involved in water and sanitation programmes – paint a stark picture. Globally, 44 per cent of household wastewater is not safely treated, severe water-related diseases such as cholera and schistosomiasis are widespread in many poor countries, and 1.8 billion people use a source of drinking water contaminated with faeces. Since the 1990s, water pollution has worsened in almost all rivers in Africa, Asia and Latin America due to industrial and agricultural waste. Currently 2.3 billion people live in water-stressed countries, of which 733 million live in high and critically water-stressed countries, and could be displaced by

---

32 See F. Genoux, 'In the midst of Chile's megadrought, anger turns toward avocados', *Le Monde*, 20 October 2022.

33 Cotton is one of the world's most water-intensive crops. See Menga, *Power and Water in Central Asia*.

2030. Rising urbanization – today 55 per cent of the world's population lives in urban areas, and the UN expect this to increase to 68 per cent by 2050 – is also putting a strain on water distribution and use.[34] Beyond the several large cities that might soon run out of water (examples include Chennai, Mexico City, Cairo and London), the rise of overcrowded slums – a process that has been thoroughly documented by Mike Davis – will add to the challenge of guaranteeing access to safe drinking water systems and adequate sanitation in the Global South.[35] As urban areas soak up much of the world's population growth in the next few decades, water infrastructure will struggle to keep pace, even though water and sanitation access rates tend to be higher in urban areas than in rural ones. As the WHO notes: 'In urban areas, the main challenge is often a lack of access to basic services in informal settlements, or high prices and a lack of quality control of water from private vendors. In rural areas, water may be free, but it may involve long journeys to and from the source, and may be contaminated.'[36] The water crisis is thus a crisis of availability, but also of access.

The aim of this book is not necessarily to provide answers; rather, it is to raise questions about the neoliberalized, pseudo-religious, technocratic solutionism deployed to 'fix' the water crisis – a solutionism that re-entrenches the very inequalities, exploitation and developmentalism it promises to overcome. The spark for this book was kindled by my initial curiosity about Water.org and the dedication of a

---

34 '68% of the world population projected to live in urban areas by 2050, says UN', UN Department of Economic and Social Affairs, available at un.org.
35 M. Davis, *Planet of Slums*, London: Verso, 2017.
36 World Health Organization, UN-*Water Global Analysis and Assessment of Sanitation and Drinking-Water (GLAAS) 2017 Report: Financing Universal Water, Sanitation and Hygiene under the Sustainable Development Goals*, Geneva, 2017, p. 56.

celebrity like Matt Damon to solving the water crisis. As a geographer who has worked on water politics for the last fifteen years, I tried to connect the dots and map this 'water network', so that I could make sense of the complex set of relations produced by the social construction of the 'global water crisis'. To do this, the book will delve into three related concepts – care, sacrifice and redemption – and link them to three interconnected case studies – the water charities Water.org and WaterAid, celebrity-led social network campaigns aimed at solving the global water crisis, and the sustainability policies of one of the largest bottled water companies in the world, Nestlé Waters. In doing so, the book not only delineates the contours and meaning of the water crisis, but also provides an analytical framework to examine the nature of charity work. It highlights the appropriation and domestication of care, sacrifice and redemption in the name of growth, progress and 'good deeds' for others. Within the book, these three concepts will be discussed rather succinctly in the interest of the book's fluency, but suggestions for further reading will be provided after the conclusions.

Chapter 1 examines in greater depth the genesis of the concept of the 'global water crisis', developing a critique of both the meaning of 'crisis' and that of 'global'. Through this, the chapter explains that while the idea of a crisis might suggest that we are experiencing something exceptional, the inflated use of the notion of a 'global water crisis' renders it ordinary and manageable. The chapter also considers the evolving choreographies of power that influence how and why water does and does not follow its natural course, focusing on the increasing influence of new and emerging transactional actors.

Chapter 2 focuses on two global water charities that are at the forefront of the global quest to solve the water crisis: Water.org and WaterAid. It traces the genealogy of global water charities to the mid-1990s and the beginning of the golden era of green neoliberalism, which was marked by contentious North–South relations and the consolidation of the water privatization agenda led by the World Bank. The chapter then outlines how global water charities are driven by an original good intention: solving the water crisis and improving the human condition. Like the Good Samaritan, they do not pass by when they see people who suffer, but instead care for them, and do this with the help of others. And yet, this original good intention gradually becomes obscured by the market logic that dominates global water governance, and the ever-pressing need to mobilize the 'real money' to fulfil their vision. A political economy of care thus emerges, assembled through partnerships with transnational water companies and investment banks, and microcredit and donation campaigns driven by consumption patterns. Through this, water charities end up reproducing, normalizing and legitimizing the same system and exploitative relations that are responsible for the inequalities and environmental problems that they try to solve.

Chapter 3 builds on the previous chapter by examining the growing influence of celebrities in global water governance. The fact that both Water.org and WaterAid are fronted by global celebrities (respectively Matt Damon and Ringo Starr, to name two) is symptomatic of a broader tendency that sees celebrities emerge as drivers and outcomes of the changing landscapes of global development. The chapter provides an overview of numerous celebrity-led campaigns run by the two water charities on social networks and fundraising platforms – including the toilet strike #strikewithme, and the 'Give it up for taps and toilets' and the #PourItForward campaigns, in which individuals are called upon to turn their self-sacrifice (donation or 'caring' purchase) into fundraising frenzies. Celebrity saviour-like Good Samaritans thus work to take both elites and fans with them into redemption through sacrifice, care and empathy monetized via donations, 'correct' purchases of 'helping' commodities and, sometimes, policy change or investments with other elite celebrities' personal wealth. Through their viral words as advocates of charity-focused neoliberal philanthrocapitalism, celebrities frame for us a thoroughly marketized and technocratic set of solutions to the water crisis and other crises bedevilling the planet and people.

Chapter 4 expands the discussion of the capitalist appropriation of the water crisis by looking at the sustainability initiatives of one of the largest bottled water companies in the world, Nestlé Waters. There are indeed several sins for which large bottled water companies seek redemption, including plastic pollution, greenhouse gas emissions, water commodification, depletion of underground aquifers, and water grabbing. Through discussion and analysis, the chapter exposes the inherent contradictions of the company's initiatives: on one hand, Nestlé Waters pledges to guarantee the 'water for all' and 'water as a human right' principles, while on the other hand, it sacrifices a tiny fraction of its revenue to this endeavour, reproducing the appearance of a real effort and keeping criticisms at bay. At the same time, the company continues its involvement in several concerning conflicts and even legal battles with local communities about water pumping and water grabbing.

The concluding chapter reflects on the current outlook of global water governance, examining how the neoliberalization of the water crisis is foreclosing the imagination and emergence of alternative socio-technical arrangements and infrastructures aimed at improving the human condition. The chapter then extends these reflections to the environmental and climate crises, considering the theoretical and empirical implications of the book for our understanding of how humanity comes to terms with socio-ecological change in the Anthropocene. The book ends by arguing that

INTRODUCTION

if humanity is to get out of its currently deadlocked set of relationships between humans and the environment, it has to reconsider its relationship with nature, one that has to be built on a renewed responsibility and a serious reconsideration of what it means to care for the planet and, inevitably, for others and ourselves.

1

Crisis Management

Making global water

In 2021 the 9th World Water Forum (WoWF) unfolded in Diamniadio, Senegal. This triannual event is organized by the World Water Council (WWC), a Marseille-based international think tank founded in 1996 and led by Loïc Fauchon, the former CEO of the Société des Eaux de Marseille, itself a private water supply company owned by the global water giant Veolia Water. The founding members of the WWC include the International Union for the Conservation of Nature (IUCN), the International Water Association (IWA), the International Water Resources Association (IWRA), the UN Development Program (UNDP), the UN Educational Scientific and Cultural Organization (UNESCO), the World Bank and the French multinational corporation Suez Lyonnaise des Eaux. The opening ceremony of the 9th WoWF brought together heads of state, CEOs and senior policymakers around the theme 'water security for peace and development'. In his welcome speech, the Senegalese president Macky Sall warned his guests that:

> There is every reason to believe that if nothing is done, the situation will get worse and worse, due in particular to strong demographic pressure, rapid urbanization, and polluting industrial activities. This 9th World Water Forum gives us the opportunity to sound the alarm on the gravity of the situation,

so that water issues remain at the heart of the international agenda.[1]

Sall paints a bleak picture, one in which there is not much reason to be optimistic. In a similar vein, Fauchon also issued a stark warning:

> Technology and the contribution of digital technology will not be enough. We must certainly innovate, innovate again, innovate always, pump, transfer, desalinate, recycle, and increase the amount of water available. But we also need to change our habits and behaviours, and only then will we be able to share water for people and water for nature.[2]

According to the WoWF, and similar to what we observed in the case of Cape Town, solving the water crisis requires innovation and technology as well as individual action. Fauchon focuses on technical-financial, behavioural and managerial changes rather than systemic change, but this should not come as a surprise.

The WoWF is one of many – and perhaps the most prominent – large water conferences regularly organized around the world. Such conferences have proliferated since the mid-1990s with the launch of the WWC, the Global Water Partnership (GWP) and the Stockholm International Water Institute (SIWI), and they have become a key component of global water governance. While these events attract policy attention to water issues, they also compete for attention, eroding the perceived legitimacy of the numerous water meetings organized by government agencies, non-profit organizations such as the IWRA, and the UN and its agencies. As Gleick and Lane explain:

---

1 L. N. Vofo Kana, 'President Sall Raises the Alarm on Major Challenge of the Century as World Water Forum Kicks Off', *Africa News*, 22 March 2022.
2 Ibid.

[T]he first World Water Forum was organized by the WWC in Marrakech, Morocco in 1997 in direct competition with the traditional IWRA triennial Congresses. This forum also set the stage for what have become triennial World Water Forums, at The Hague in 2000, in Kyoto in 2003, and in Mexico City in 2006. Professional attendance at these Forums has grown from 500 at Marrakech through 5,000 at The Hague to 12,000 to 24,000 at Kyoto.[3]

Large water congresses tend to have a broad focus and scope, and thus reiterate a decontextualized sense of urgency about solving the 'global water crisis' without, however, providing specific or concrete measures for doing so. It is hard to disagree, for example, with the main principles guiding the Dakar Declaration,[4] the text that sealed the 9th WoWF, calling for a 'blue deal for water security and sanitation for peace and development'.[5] The document is signed by the stakeholders of the event – a range of multinational companies, national governments and some UN agencies – and calls upon the international community to guarantee the right to water and sanitation for all, enhance cooperation and ensure inclusive water governance.

---

3 And to over 30,000 attendees at the five most recent editions. P. H. Gleick and J. Lane, 'Large International Water Meetings: Time for a Reappraisal. A Water Forum Contribution', *Water International* 30: 3 (2005), p. 411.

4 Ironically, and despite its name, the Dakar Declaration was not adopted in Dakar, but rather in the nearby town of Diamniadio. On top of this, the mayor of Dakar, Barthélemy Toye Dias, boycotted the forum and ran the alternative event 'L'eau à Dakar, Dakar dans les eaux' (Water in Dakar, Dakar into waters), feeling that his city had been de facto left at the margins of the event and could not play any meaningful role in its organization. M. Ficou, 'La ville de Dakar boycotte le 9ème Forum mondial de l'eau et annonce un contre-Forum', *VivAfrik*, 17 March 2022.

5 *Dakar Declaration: A 'Blue Deal' for Water Security and Sanitation for Peace and Development*, 9th World Water Forum, Dakar, 2022.

This is in line with the content and tone of previous declarations released at the end of each WoWF, which tend to present an urgent call for decisive action on water, regularly declaring that the 'time to act is now'. The slogan of the 6th WoWF, held in Marseille in 2012, was indeed 'Time for solutions', and its webpage prominently features a pledge from Pauline, a forum participant, who promises 'to take a shower instead of a bath'. But if the need to act is so urgent, and if, each time that they are made, these calls for action sound more pressing, earnest and seemingly multipartite, how can we explain that at least 2 billion people are still using contaminated drinking water, and that inequalities between the richest and poorest in basic service coverage have increased between the years 2000 and 2017?

The reach of this paradox goes well beyond the domain of water governance. Consider, for instance, how the ecological anxieties raised by the Anthropocene so often culminate in predictions of total apocalypse. By way of example, we might think about the Climate Clock, an online stopwatch and artistic installation that has appeared in several cities around the world. The Climate Clock displays a countdown, ending in early 2028, when 'famine, drought, floods, displacement, conflict, suffering and disaster' will be unavoidable unless human societies drastically reduce their CO_2 emissions to keep global warming beneath the 1.5°C threshold.[6] The clock's insistent ticking is designed to remind humanity that the 'Earth has a deadline', stirring the same anxieties that circulated ahead of the anti-climactic Y2K countdown. The very notion of a 'deadline' has a particularly managerial ring to it, and the linear movement it puts into play, characterized as a relentless forward march, is echoed in numerous policy and advocacy platforms from across the political spectrum.

---

6 For further information refer to the Climate Clock website (climateclock.world).

My point here is *not* to deny the veracity of these many warnings, premonitions and ever-tightening deadlines. It is to highlight, echoing commentators such as David Wallace-Wells, that global CO_2 emissions have continued to grow – indeed, to increase exponentially – regardless of, and at the same time as, these stark and increasingly apocalyptic predictions.[7] While it is true that emissions growth has finally slowed over the last few years, it is clear that it is still yet to reach its peak. It can be suggested, with a degree of self-conscious polemicism, that while the messianic environmentalism typified by forewarnings against 'the excessive predations of a beastly humanity' may be anxiety inducing, they have so far failed to bring about the change for which they advocate, nor have they inspired sufficient, globally coordinated, political action.[8] This does not mean that they have been wholly ineffective; rather, that these effects have sometimes been of a very unintended kind. Most notably, warnings of singular catastrophic moments and unsurpassable deadlines have successfully transformed a dominant understanding of nature and of human relationships with the environment into a *spectacle*. This spectacularization of the climate – and water – crisis fits well into the conceptual rubrics and economic rhythms of capital, absolving and evolving – rather than identifying – the system that has caused the climate crisis in the first place.

But let us return to the WoWF. I was in Marseille in 2012 for its sixth iteration, coming 'home' to the city that headquarters the WWC. I thus paid its steep registration fee (a weekly pass costs around €400, a little less if you are a student) and embarked on a ferry from Sardinia to Marseille. It was March, and after a day of meetings and interviews

---

7 D. Wallace-Wells, *The Uninhabitable Earth: Life After Warming*, New York: Columbia University Press, 2020.
8 A. Badiou, 'The Neolithic, Capitalism, and Communism', trans. David Broder, versobooks.com, 2018.

with former UN colleagues and water experts, I decided to go for a walk around the Vieux Port. As soon as I arrived there, I noticed a large and festive group of people that was marching towards me from la Canebière. Some of them were singing and playing drums, and I could see flags and banners in several languages, reading 'L'eau source de vie, pas de profit', 'Water for people not for profit', 'Si scrive acqua, si legge democrazia' or 'El agua como la vida no es una mercancia'. It was the Alternative World Water Forum (FAME 2012, from the French 'Forum Alternatif Mondial de l'Eau'). As stated on its webpage, the aim of the FAME is 'to create a concrete alternative to the World Water Forum which is organized by the WWC. This Council is a mouthpiece for transnational companies and the World Bank and they falsely claim to head the global governance of water.' I talked to some of the activists, and most of them told me that they were there to protest against the perceived illegitimacy of the WoWF: its mandate does not originate from communities or their governments but is rather the expression of the will of some very large multinational companies.

Indeed, as I attended the WoWF closing ceremony, I could not help being slightly confused. The organizers showcased a series of ministerial declarations 'for the future of water on our planet and for a sustainable development'. These declarations and commitments were being uttered not only by government ministers and deputy ministers, members of national parliaments, but also by CEOs of large private companies and other stakeholders. There was a sort of 'UN' taste to it, but the WWC, and not the UN, was running the event. The WoWF can hardly be considered inclusive or democratic. Participation is subject to a heavy fee, particularly for exhibitors, and not all countries, companies or NGOs are invited or can afford to be there. Since the WoWF is attended predominantly by large multinational companies who are there because of their economic power and are only accountable to their shareholders, there is no reason to believe that they will prioritize the needs of the world's poorest over those of their shareholders. And let's be clear: this is neither illegal nor surprising; it is systemic.

Even more so, the WoWF, and with it the notion of a 'global water crisis', is underpinned by a straightforward narrative that tends to juxtapose two main categories: those who have water, and those who don't. And inevitably, those who do are supposed to help those who don't. Consider, for example, the speech given by the president of France, Nicolas Sarkozy, ahead of the Marseille WoWF:

> The present moment remains, however, a serious moment: because the question of water is one of the heaviest issues of the coming century. The abundant and healthy water flowing from every tap in our countries should not make us forget that, at this very moment, in many parts of the world, water is dirty or too scarce. Every fifteen seconds, a child dies in the world because of dirty water. In today's world, water kills more people than wars, famine, or AIDS ... The figures overwhelm us: 1 billion people do not have access to drinking water; 2.5 billion people do not have access to sanitation; 8 million people, including 2 million children, die every year from the unsafe water they drink ... This is why France has, since 2005, doubled the funding granted for water and its sanitation under international public aid. The state is not alone in this cooperation. Thanks to the law of 2005, known as Oudin-Santini relating to decentralized cooperation, local authorities and water agencies can carry out their own solidarity actions with towns and villages in the South.[9]

---

9 *Déclaration de M. Nicolas Sarkozy, Président de la République, sur le 6ème Forum mondial de l'Eau organisé à Marseille en 2012*, à Paris le 2 juin 2010, available at elysee.fr. Original French version translated by the author.

Sarkozy's binary thinking is evident – his world is divided into two main groups – and with it comes an overall securitization of water and the issues that surround it, thus joining the by-now insistent calls in media and policy circles about an impending 'water war' and conflicts over 'blue gold'. Two aspects of Sarkozy's speech are also significant. The first is the mention of 'Oudin-Santini', a law that allows French local governments and water-basin agencies to devote up to 1 per cent of their water and sanitation budget to aid and development projects overseas. What is interesting here is that the law is based on the interconnection between consumption and solidarity: the more water is consumed, the more money becomes available for development initiatives.[10] But are we sure that this is the best way to bring water for all, or are we rather in front of a perverse incentive, encouraging, directly or indirectly, greater resource use, thus unintentionally leading to the degradation of that same resource? This brings us to the second aspect, and that relates to how these development initiatives are actually unfolding on the ground.

The biggest European water company is France's Veolia, which in 2022 bought what was at the time the second biggest company, Suez, which is also French. Veolia – together with what was formerly known as Suez – has a market value of over US$40 billion, and its declared goal is to manage the entire water cycle. Veolia-Suez is deeply involved in the WoWF, and is also very active in Africa, where most French development initiatives also take place. France controlled a significant colonial empire in Africa, and even though its formal rule ended in 1962, the country still has considerable interests in the continent. Suez, for example, has built over 500 water plants in Africa and manages several private water systems across the continent, including the drinking water

---

10 A. Bousquet, *Support Mechanisms for WOPs from the North: The Oudin-Santini Law and Decentralized Cooperation in France and Other Examples*, December 2009.

and sanitation services of Greater Casablanca and Algiers in the former French colonies of Morocco and Algeria. The company has also been involved in serious controversies over water distribution and use, including the installation of pre-paid water meters – a system that is considered illegal or unconstitutional in several countries – in Johannesburg in the early 2000s, causing significant protests. In 2019, as preparations for the Dakar WoWF were already underway, Suez was awarded a contract for producing and distributing drinking water in Senegal (also a former French colony). Suez's celebratory press release is telling of the company's largely neo-colonial attitude and of its political economic implications:

> Through this contract, the Senegalese government is seeking to harness the expertise of the Suez Group to meet the drinking water needs of the country's quickly growing population. Sufficient access to high-quality water is a key issue in Senegal and a priority in the Emerging Senegal Plan launched by President Macky Sall.[11]

The underlying assumption is that Senegal and its government are underdeveloped, possibly corrupt and thus are not able to efficiently manage their water system. They therefore need to hand over this responsibility to a private company which 'knows better' and can be trusted to protect the interests of Senegalese citizens, together with those of Suez's shareholders. This, of course, disavows centuries of deliberate exploitation of Africa at the hands of the European powers, a process brilliantly narrated by Walter Rodney in his book *How Europe Underdeveloped Africa*. Just as Europe underdeveloped Africa, Africa developed Europe, and this applies also to infrastructure, companies and whole

---

11 SUEZ, *Suez Was Awarded the Contract for Producing and Distributing Drinking Water in Senegal*, Suez press release, 30 December 2019.

industrial sectors. Rodney specifically references Senegal's capital Dakar and the workings of the French fund for the development of its colonies, FIDES:

> The high proportion of the 'development' funds went into the colonies in the form of loans for ports, railways, electric power plants, waterworks, engineering workshops, and warehouses, which were necessary for more efficient exploitation in the long run. In the short run, such construction works provided outlets for European steel, concrete, electrical machinery, and railroad rolling stock. One-fifth of FIDES funds were spent on prestigious public works in Dakar, which suited French industry and employed large numbers of expatriates. Even the schools built under FIDES funds were of unnecessary high cost per unit because they had to be of the requisite standard to provide job outlets for white expatriates. Incidentally, loans were 'tied' in such a way that the money had to be spent on buying materials manufactured in the relevant metropole.[12]

In the past, colonial actors looked to modernize Africa by introducing some elements of capitalism to the continent (such as private property in land and private ownership of other means of production). In the twenty-first century, Suez – and with it, the French government – is still trying to modernize the continent and its water infrastructure. This time, they are doing so by introducing some key features of contemporary capitalism, including market-driven water privatization, water capture and an increasingly promiscuous relationship between the public and the private spheres. It is clear that, while at the supply level water is an essentially local resource, there are numerous global and systemic forces that determine how the resource is supplied, and what kind of investments are made (or not made) in the water sector.

---

12 W. Rodney, *How Europe Underdeveloped Africa*, London: Verso, 2018, pp. 257–8.

By the same reasoning, the ways in which the WoWF and the other main players in global water governance present themselves to both the public and the market might suggest that they are under a global mandate to solve a seemingly global problem, but, as we shall see, this is far from being true.

Whose water crisis?

A crisis is, by definition, a temporary situation, a decisive moment and 'an unstable or crucial time or state of affairs in which a decisive change is impending'.[13] But the current water crisis is hardly temporary, and there is no decisive change on the horizon. As it is currently employed, the concept of the 'global water crisis' refers to a quite stable, almost reassuring state of things: there was a water crisis when most of us were born, there is one now and there will be one when we will be gone. So, while the idea of a crisis might suggest that we are going through something quite exceptional, the inflated use of the notion of a 'global water crisis' renders the crisis ordinary. This does not come as a surprise: as we have already noted, crisis – and with it, profit-making crisis management – is one of the structural foundations of capitalism. So, when did we start talking about a 'global water crisis', and what does that mean? The first recorded use of the term dates back to 1961, when *Health*, the Australian journal of the Commonwealth Department of Health, wrote:

> The world is on the verge of a global water crisis because of increasing demand and increasing pollution. About 1,600 million people were alive in 1900; over 2,000 million in 1930: about 3,000 million in 1960; and it is estimated that by the year 2000 the world's population will amount to over 6,000 million people using 1,000 gallons of water for every ton of

---

13 Merriam-Webster Online Dictionary.

bread, 5,500 gallons for every ton of butter, 71,000 gallons for every ton of steel.[14]

Apart from the dates and figures provided, the above text could have easily been written today, since much of the contemporary thinking around the 'global water crisis' is still largely influenced by basic neo-Malthusian assumptions about human population and its water footprint. In 1969, Georg Borgstrom made similar claims in his book *Too Many: A Study of Earth's Biological Limitations*.[15] Also of note, in 1969 Raymond Nace, a hydrologist, wrote the report *Water and Man: A World View* for UNESCO, which provided a universal overview of the co-constitutive relationship between societies and hydrology.[16] But it is in 1977, with the UN Water Conference in Mar Del Plata, Argentina, that the concept gained traction. This conference was the first of its kind, and it was attended by over a hundred governments and several international and non-governmental organizations. The event came as a response to a series of important water crises that occurred in the 1970s, including severe droughts in the Sahel, Western Europe and North America, and followed other UN conferences on so-called global concerns, such as the environment (1972, in Stockholm), food (1974, Rome) and human settlements (1976, Vancouver). It also took place following the initial stimulus provided by the Club of Rome and its 1972 report *The Limits to Growth*. The formation of the Club of Rome in the late 1960s marked a key moment in the resurgence of Malthusian theory while also initiating the debate on 'global problems' and how to solve them. Matthias Schmelzer has noted that there is

---

14 *Health: Journal of the Commonwealth Department of Health*, 1961, p. 106.

15 G. Borgstrom, *Too Many: A Study of Earth's Biological Limitations*, London/Toronto: Macmillan, 1969.

16 R. L. Nace, *Water and Man: A World View*, Paris: UNESCO, 1969.

a significant overlap between the key people behind the Club of Rome and the Organization for Economic Co-operation and Development (OECD). These were highly educated white men from the Global North, mostly with backgrounds in the natural sciences and engineering, acting as philanthropists from economic positions of privilege and power, including Aurelio Peccei (a philanthropist and former high-level manager), Alexander King (the Director-General for Scientific Affairs at the OECD) and Hugo Thiemann (a research and development manager who worked for the Nestlé Group). As Schmelzer notes:

> King and Peccei immediately discovered that they shared a 'vision of global dangers that could threaten mankind such as over-population, environmental degradation, worldwide poverty and misuse of technology'. Since there 'did not seem to be any single body capable of analyzing, let alone starting significant action against the global threats', they sat down in King's office in the OECD's headquarters, the Château de la Muette, and drew up a list of people whom they wanted to involve in these issues.[17]

Two aspects stand out here: the first is these men's attention to *global* issues, and the second is their shared feeling of existential insecurity. Following a few preparatory meetings, the Club of Rome was founded in 1968, and soon after it commissioned *The Limits to Growth* report. The report, which is dedicated to Peccei – 'the prime moving force within the group' – and his project 'on the predicament of mankind', made use of computer simulations to develop a critique of the growing, and unsustainable, footprint of humanity. It is, to date, the best-selling environmental book in

---

17 M. Schmelzer, '"Born in the Corridors of the OECD": The Forgotten Origins of the Club of Rome, Transnational Networks, and the 1970s in Global History', *Journal of Global History* 12: 1 (2017), p. 34.

history, with over 30 million copies sold worldwide, and this work is arguably responsible for both our current concern over sustainable development and our widespread anxieties about overpopulation and environmental scarcity. But more than this, the establishment of the Club of Rome in 1968 is a breakthrough moment in the emergence of intellectual Atlanticism, and the consolidation of the transnational networks and expertise that inform the contemporary understanding of the 'global water crisis'. While these transnational networks bring together representatives of both the private and the public sectors, they also mark the rise of techno-optimism and private-led initiatives in international development, and the consequent erosion of the power of the state. Most mainstream studies of international politics and geopolitics suggest that the end of the Cold War in 1991 also coincided with the emergence of the environmental crisis – rather than military violence – as the key international issue of our times, and yet it is clear that this process began much earlier. The current dominance of environmental forecasting and climate modelling – and our obsession with the looming apocalypse: 'There is only ten years left for humanity to act before it's too late!' – can also be traced back to the Club of Rome and the ideas developed therein. Take, for instance, these words by Peccei from the early 1970s:

> I believe that we must have the courage of utopia. If we look closely at the future of man, of humanity, of our children and our children's children, it is something that we forge and that we imagine or that we unintentionally clarify ourselves. The future, in this human sense, is a social invention of the community and therefore we must have the courage to see a better future. The best that can be obtained in the real conditions of today and those that will exist tomorrow. Therefore, speaking of utopia with mockery or scepticism seems to me to do nothing but put us in an even more difficult condition than we are in now because

we demonstrate that we do not have the will, the courage, the vision to look for something better in the future than what we have today on this earth. Today is perhaps the last chance we have, but if we don't change course, if we continue to do what we are doing more or less now, all together, we will surely head towards a precipice. If we continue on the current road there is no doubt that we are heading towards much greater crises, probably towards disasters, probably towards catastrophes.[18]

While today most billionaires act as illuminated visionaries on a mission to save the world – think of Jeff Bezos, Elon Musk, Bill Gates or Richard Branson – Aurelio Peccei was doing exactly the same before it was trendy. Peccei is possibly the father of all visionaries, the visionary *ante litteram*, and even though he is largely unknown to most, his influence on mainstream thinking about the socioecological crisis deserves to be studied further. But, for our purposes here, it is sufficient to note that the ideas behind the Club of Rome have also contributed to shaping the understanding of the water crisis as a global one. This interpretation is misleading, as we shall see in a moment.

In any case, the late 1970s offered ideal conditions for the circulation of the concept of the global water crisis among political and economic elites. Following the 1977 UN Water Conference, things escalated rapidly. In 1980, the UN General Assembly proclaimed the Declaration of the International Drinking Water Supply and Sanitation Decade, and in 1981 the British National Water Council organized the Thirsty Third World (TTW) conference, which eventually led to the birth of WaterAid. It was also during these years, as Jeremy Schmidt observes, that the International Water Resources Association (IWRA) – an international

---

18 *Documentario commemorativo 50° Anniversario de I Limiti della Crescita e Aurelio Peccei*, available at YouTube.com. Original Italian version translated by the author.

organization established in 1971 – began to exert a growing influence on global water governance.[19] The IWRA held the first of many World Water Congresses in 1973 in Chicago, and in 1975 it launched the journal *Water International*. At the same time, the IWRA partnered with global organizations (including UNESCO, UN-Habitat and the World Bank) to position global water governance at the point at which water's global distribution intersects with human needs. As Schmidt explains, during the Mar del Plata conference these two aspects – global water distribution and human needs – provided the foundations 'for a new discourse of water scarcity that ported the scientific objectivity of global hydrology over to social policy'.[20] Indeed, the often-cited statistic reporting that only 0.1 per cent of the world's water is held in freshwater lakes and rivers contributed to the diffusion of rational planning; in other words, the utilitarian idea that scarce water resources ought to be measured and quantified, and subsequently managed rationally to ensure human well-being, something that the World Bank keenly translated into the need for water pricing.

Thus, in the 1980s, when neoliberalism spread its wings, a watery transnational (or global) platform composed of private institutions, international organizations, think tanks, the UN and the World Bank was already in place and ready to adopt its precepts to solve the water crisis: privatization, deregulation, marketization and valuation. In 1992, the Dublin Conference – a preparatory meeting of the UN Conference on Environment and Development (UNCED) – established four guiding principles for managing freshwater resources, known as the Dublin principles: i) fresh water is a finite and vulnerable resource, essential

---

19 J. Schmidt, *Global Social Policy at the Nexus of Water, Energy and Food*, in N. Yeater and C. Holden (eds), *Understanding Global Social Policy*, Bristol: Bristol University Press, 2022, p. 284.

20 Ibid.

to sustain life, development and the environment; ii) water development and management should be based on a participatory approach, involving users, planners and policymakers at all levels; iii) women play a central part in the provision, management and safeguarding of water; iv) water has an economic value in all its competing uses and should be recognized as an economic good. Of the four principles, the fourth is clearly the odd one out, and it was used in the years to follow, up to the present day, to back measures that lead to water privatization and capture. The 1990s also saw the birth of other international organizations due to become very influential in shaping the discourse of global water governance, including the SIWI (1991), the GWP (1996) and the WWC (1996), which runs the WoWF.

We then arrive in the twenty-first century. The environment – and with it, water – emerged as a key national security issue of the time, and the water wars narrative gained momentum. Realist scholars and sensationalist media articles contributed to spread the narrative, which was then adopted by leading politicians and former UN secretary-generals, who also warned about an imminent water war. In 2003, Boutros Boutros-Ghali predicted that 'the next war in the Middle East will be fought over water, not politics', and in 2011 Kofi Annan added that 'fierce competition for fresh water may well become a source of conflict and wars in the future'. Ban Ki-moon also underlined the potential of water to fuel wars and conflicts, explaining that 'water scarcity threatens economic and social gains and is a potent fuel for wars and conflict'.[21] And even though water wars never happened, the ever-expanding popularity of water scarcity scenarios and modelling inevitably resulted in the publication of numerous reports that keenly identified the main water scarcity and conflict hot spots around the world. This was done by

---

21 Menga, *Power and Water in Central Asia*.

all those involved in the transnational networks of global water governance: international organizations, think tanks and water charities (as we shall see in the following chapters), governments and intelligence agencies, and also private companies and the market.[22]

Today, the concept of the 'global water crisis' is thus used widely, and most of the time this is done without discerning its real meaning. The notion has become a container for a range of issues related to water, and it loosely refers to the combined effects of: i) water scarcity; ii) poor water governance; iii) excessive water withdrawals from agriculture; and iv) issues related to overpopulation. This interpretation provides a largely apolitical reading of the many water crises currently occurring across the world, as it is devoid of any reference to unequal water allocations across communities, water grabbing and enclosure. Consider, for instance, the research on plumbing poverty conducted by Shiloh Deitz and Katie Meehan. Their work has shown that almost half a million American households – particularly in rich cities like San Francisco and Portland – lack basic indoor plumbing, and thus do not have access to clean and safe water.[23] None of this is part of the common interpretation of the 'global water crisis'.

Of course, the mainstream framing of the 'global water crisis' narrows down the possible set of solutions deployed to fix it, which essentially focus on greater efficiency, technological improvements and rational use, and tend to disregard issues related to environmental and social justice. The poor

---

22 Refer for instance to the 2021 report *The Future of Water: Water Insecurity Threatening Global Economic Growth, Political Stability* prepared by the US National Intelligence Council's Strategic Futures Group.

23 S. Deitz and K. Meehan, 'Plumbing Poverty: Mapping Hot Spots of Racial and Geographic Inequality in us Household Water Insecurity', *Annals of the American Association of Geographers* 109: 4 (2019), pp. 1092–109.

and the marginalized are not part of the 'global water crisis'; their voices have been silenced, and yet they are the ones who really suffer from it.

Managing global water: commons and commodities

The ideological and cultural patterns that I have just discussed have, of course, many practical effects on how water is managed and distributed. Urban political ecologists have long critiqued the neoliberalization of the water crisis and the processes of commodification of water that shape water flows and, ultimately, work to marginalize poor urban dwellers globally. Overall, the commodification of water and the changing institutions and governance arrangements in the delivery of water services reflect a more fundamental shift in state–society relations, whereby crucial state responsibilities are increasingly delegated to non-state actors. The rise of non-state actors has involved two interrelated processes, encompassing a shift from vertical state-centred to horizontal governing arrangements and the simultaneous globalization of water governance challenges. Throughout much of the twentieth century, state-led programmes, driven by Fordist-Keynesian ideals, aimed to strengthen welfare states through, among others, large scale infrastructure investments, subsidization of basic services and a focus on universal access and social equity. In line with this logic and these goals, water services were valued and managed as commons. While there is no single definition of a commons, the fundamental principle is that resources like water, which are crucial to the very existence of humans, are to be considered a common good and are to be managed accordingly. To be clear, urban water services do not have a history of communal practices and institutions – governance beyond the market and beyond the state, based on (among others) 'self-determination of commoners in managing their shared resources' with reciprocity and collective decision-making – as is the case for forest or groundwater resources.[24] However, as Maude Barlow – a Canadian author who argues for the re-municipalization of water and sanitation services – explains, the concept of universal access grounded in the 'membership' of humanity has been extended to 'social commons' like water services.[25] Based on this logic, water services were owned and controlled by the state and managed as a collective consumption good, regulated and provided by a public body.

Commons, however, pose a profound challenge to capitalist logics, as they are reliant on communal relations and properties that capitalism needs to eradicate in order to grow and proliferate. The 1970s marked a departure from the collective consumption good in governance discourses, and in its place appeared a new logic of public water services management as unable to deliver basic services to the growing urban population, driven by the emerging discourse of the overloaded state. In the past, increased public expenditure had been seen as the most effective approach by which to expand water coverage to unserved or underserved citizens. By contrast, the emerging discourse of 'state failure' portrayed states (and particularly former colonies, as I discussed earlier in the case of Senegal) as both ineffective and inefficient in delivering water services and ensuring environmental protection of the resource. Framings of 'state failures' have carefully overlooked significant achievements in ensuring universal access to basic services in the Global

---

24 P. Linebaugh, *The Magna Carta Manifesto: Liberty and Commons for All*, Berkeley: University of California Press, 2008, p. 396.

25 M. Barlow, 'The Growing Movement to Protect the Global Water Commons', *Brown Journal of World Affairs* 17: 1, Fall/Winter 2010, pp. 181–95.

North and have effectively served to mainstream the notion of water as an economic good (as opposed to a merit or a quasi-public good) and to promote marketization logics. Mainstream neoclassical economists and conservatives – like James Winpenny, for instance – linked the assumed inability of the state to supply water to its growing urban population to 'the failure to treat water as a scarce commodity'.[26] Accordingly, and in line with the fourth Dublin principle, valuing water as an economic good is considered necessary to manage it efficiently and to ensure water conservation. Water services are thus subsumed in tradable and profitable commodities, and water scarcity is discursively framed as both a natural phenomenon to be addressed through full-cost recovery tariffs and a social condition generated by state failures. These arguments provide additional justification for privatization, private sector participation and the commercialization of water services.

This fundamental ideological shift ultimately embodies what Karen Bakker defined as 'market environmentalism'; in other words, an attempt to reconcile conservation, efficiency and economic growth.[27] This approach has had significant implications for both the institutions that govern service provision and the organizations tasked with this responsibility. As a consequence, the late 1980s and 1990s were characterized by a process of (partial) 'hollowing out' of the state, whereby the delivery of basic services was incrementally outsourced to non-state actors, including the third sector (NGOs, community enterprises, etc.) and private companies. Increasingly, therefore, governments relied on hybrid and horizontal forms of governing, in which a wide spectrum

---

26 J. Winpenny, *Managing Water as An Economic Resource*, London: Routledge, 1994, p. 1.

27 K. Bakker, 'The "Commons" versus the "Commodity": Alter-Globalization, Anti-Privatization and the Human Right to Water in the Global South', *Antipode* 39: 3 (2007), pp. 430–55.

of actors was involved in the delivery of basic services. This 'governance beyond-the-state', as Erik Swyngedouw called it, refers to the shifting relationship between state and non-state actors, characterized by:

> The emergence, proliferation and active encouragement ... of institutional arrangements of 'governing', which give a much greater role in policy-making, administration and implementation to private economic actors on the one hand and to parts of civil society on the other hand in self-managing what until recently was provided or organized by the national or local state.[28]

Concurrently, institutional reforms have reframed water as an economic good: principles of economic equity replaced social equity (respectively, the 'willingness-to-pay' and the 'ability-to-pay' principles) in water pricing, and a new 'hydrosocial contract' emerged, which involves consumers (rather than citizens) and is founded on the elimination of subsidies and the conception of water as a quasi-commodity. Proponents of this approach claim that this will warrant more inclusive water services by ensuring efficient demand management, strengthening utilities' performance and 'empowering' consumers, thereby ensuring service expansion to unserved or underserved areas. Driven by these ideas, financial institutions such as the International Monetary Fund (IMF) and the World Bank started to impose domestic water privatization policies as a pre-condition for a loan to highly indebted poor countries. As a result, by the early 2000s, water services in over 100 cities were managed by a few multinationals, which saw service delivery as an emerging arena for profit and capital accumulation. Michael Goldman notes that as recently as 1990, fewer than 51 million people received their water from private water

---

28 E. Swyngedouw, 'Governance Innovation and the Citizen: The Janus Face of Governance-beyond-the-State', *Urban Studies* 42: 11 (2005), p. 1992.

companies, primarily in Europe and the United States. Ten years later, that number had risen to 460 million, and the highest growth rates were recorded in Africa, Asia and Latin America.[29] Today, AquaFed – the International Federation of Private Water Operators – estimates that around 800 million people are served by private operators, and we can of course expect this number to continue to grow. Local commons are thus dispossessed and turned into global capital, which is in turn transferred to capital-thirsty global markets.

Several UN-led initiatives and global campaigns keep bringing together governments, international and non-governmental organizations, and private companies to develop the technical, institutional and financial capacity needed to increase service levels and expand coverage in the Global South. These events include the UN International Year of Freshwater (2003), the 2005–2015 UN International Decade for Action on 'Water for Life', the Millennium Development Goals' (MDGs) pledge to halve the proportion of people without access to water by 2015, the International Decade for Action 'Water for Sustainable Development' 2018–2028, alongside the standalone goal of universal access to water and sanitation (goal 6) in the SDGs. The UN Agenda 2030 also places water as a central development objective. However, these policy instruments and development targets have not generated the transformative ideological and institutional change needed to progressively eradicate urban water inequalities. On the contrary, these programmes were largely designed around the notion of water as an economic good, and water sector reforms promoting this principle were imposed to low- and middle-income countries which depend on financial resources disbursed by Global North–dominated lending agencies

---

29 M. Goldman, 'How "Water for All!" Policy Became Hegemonic: The Power of the World Bank and Its Transnational Policy Networks', *Geoforum* 38: 5 (2007), pp. 786–800.

like the World Bank and the International Monetary Fund (IMF). Maria Rusca and Klaas Schwartz have identified three interrelated, and at times overlapping, strategies of water commodification. During the 'privatization decade' (1993–2003), large multinationals – assumed to have comparative advantages in accessing financial resources and promoting performance based management, flexibility and efficiency – began privatizing water services sectors in the Global South.[30] As water services are an 'uncooperative' commodity (subject to market failure), private companies have increasingly focused on risk reduction by withdrawing from non-profitable contracts and moving to less risky forms of participation, such as water operator partnerships. This mode of water commodification has been followed (and partially replaced) by the commercialization of public water utilities through the introduction of commercial principles (for example, financial sustainability and full cost recovery). In parallel, small-scale water providers, previously dismissed as 'informal' or 'illegal', are increasingly promoted as an effective complement to utilities. Their reappraisal has been largely grounded in their private sector–like characteristics, including their ability to invest, recover costs, operate without subsidies, and their efficiency.

The implementation of these reforms has had profound implications for the urban water services sector in the Global South. Today, in most cities the orthodoxy of full cost recovery means that water utilities only supply between 40 and 70 per cent of the urban population, while the rest – often the most vulnerable and low-income dwellers residing in informal settlements, as we have seen in the case of Cape Town – must rely on self-supply, on an array of small-scale

---

30 M. Rusca and K. Schwartz, 'The Paradox of Cost Recovery in Heterogeneous Municipal Water Supply Systems: Ensuring Inclusiveness or Exacerbating Inequalities?', *Habitat International* 73 (2018), pp. 101–8.

water providers operating as social enterprises and on market-led water charities. The proliferation of the latter is indeed both a product of, and a response to, the abovementioned shifts in the global water governance agenda, and further exemplifies the evolving nature of capitalism and its ability to continuously expand into new territories. Capitalism has indeed absorbed the water crisis, which is, in turn, the result of its own workings.

The spirit of capitalism

To understand what is happening in global water governance, I will now take a small detour to discuss the spirit of capitalism and its manipulation of conceptions of care, sacrifice and redemption. To do so, I suggest turning to Walter Benjamin's essay 'Capitalism as Religion', which, despite being written in 1921, remains deeply relevant in our time. In the essay Benjamin maintains that '[a] religion may be discerned in capitalism – that is to say, capitalism serves essentially to allay the same anxieties, torments, and disturbances to which the so-called religions offered answers'.[31] While Benjamin was clearly inspired by Max Weber's *Protestant Ethic and the Spirit of Capitalism*, unlike Weber, he did not interpret the development of capitalism as a consequence of the secularization caused by Protestantism. Rather, he saw it as a parasite of Christianity, to the extent that in modernity, 'Christianity's history is essentially that of its parasite – that is to say, of capitalism'.[32] Three aspects, according to Benjamin, underpin this religion of modernity, with religion understood in Durkheimian terms as a shared

---

31 M. Bullock and M. W. Jennings, *Walter Benjamin: Selected Writings, Volume 1: 1913–1926*, Cambridge, MA: The Belknap of Harvard University Press, 2016, p. 288.

32 Ibid., p. 290.

system of beliefs that unite a given group: i) capitalism is a cultic religion, where everything has meaning only in relation to the cult; ii) this cult is permanent, as there is no difference between working days and feast days; and iii) this cult is based on a universal guilt. For Benjamin, guilt is a driving and pervasive force in capitalism, 'to the point where God, too, finally takes on the entire burden of guilt, to the point where the universe has been taken over by that despair which is actually its secret *hope*'.[33]

As capitalism tends towards guilt and despair, its aim, as a religion, is not to transform the world but instead to destroy it. Building on the work of Benjamin and centring his critique on Nixon's decision to suspend the convertibility of the dollar into gold in the 1970s, Giorgio Agamben argues that capitalism is a religion in which faith – a word that is synonymous with *credito*, the past participle of the Latin verb *credere* (to believe) – has replaced God.[34] How can we otherwise explain the fact that today money is fully detached from its materiality (for example, gold), becoming an object whose status depends on the fact that we *believe* in its immaterial and unperishable value? Indeed, as the purest form of credit is money, capitalism is a religion in which money is the God, and banks, through their ability to produce and govern credit, have replaced churches and are now the managers of faith. But a society that increasingly relies on credit will live on credit, whereby corporate capital is increasingly sustained by a fictitious monetary capital. The capitalist religion is thus sustained by an act of faith in a future income that will help reduce this foundational debt, and banks emerge as the priests that administer the sacrament of credit debt.

---

33 Ibid., p. 289. Emphasis in original.
34 G. Agamben, (*Creazione e Anarchia. L'opera nell'Età della Religione Capitalista*) *Creation and Anarchy: The Work of Art in the Age of Capitalist Religion*, Torino: Neri Pozza, 2017.

The extract from Benjamin's 1921 piece is the basis for a crucial understanding of modernity, which, as Michael Löwy explains, transforms Weber's 'value-free' analysis of the Calvinist/Protestant treason of the true spirit of Christianity 'into a ferocious anticapitalist argument'.[35] Consistent with Marx's view of capitalism, modernity is not characterized by disenchantment, rather by the affirmation of a new religion: the transformation of the Christian spirit into the spirit of capitalism. As Jacques Lacan observes, one of the great achievements of capitalism is its ability to exploit subjective individual desire and transform it into an object that can be produced, bought and consumed.[36] Through the industrialization of desire, the hedonistic spirit of capitalism replaces the Christian spirit from which it originated. The god Mammon, the embodiment of material wealth in the New Testament (paralleled by the god Plutus in Greek mythology) and its countless manifestations – banknotes, bonds, credit and so on – is the highest deity in capitalism, an intensely social religion driven by growth to the extent that the cult of growth has fully colonized Western imagination. More precisely, as Serge Latouche observes, two phenomena – the cult of embodied value and faith in progress, technology, science and indeed growth – have led us to idolatrize the market instead of the 'golden calf'.[37] And as Jean Baudrillard remarked, as we attach value to a dematerialized entity such as money, consumables and consumption emerge as new forms of the sacred that help individuals improve their social standing, encouraging social stratification. In this interpretation, Marxian use values and exchange

---

35 M. Löwy, 'Capitalism as Religion: Walter Benjamin and Max Weber', *Historical Materialism* 17: 1 (2017), p. 60.
36 J. Lacan, *Lacan in Italia, 1953–1978*, Rome: La Salamandra, 1978.
37 S. Latouche, 'The Golden Calf Vanquishes God: Essay on the Religion of Economics', *Revue du MAUSS* 1 (2006), pp. 307–21.

values of consumables are replaced by their sign value, in that what essentially counts is not the possession of an object but our endless desire to acquire it, by our 'consumption of consumption' in a society in which the 'only objective reality of consumption is the idea of consumption'.[38]

This leads us to an important node in the development of this analysis: that of the fetish. The notion of the 'fetish' was first used in the fifteenth century by Portuguese sailors and traders travelling along the West African coast, and it refers to amulets and objects believed to have supernatural, almost magical, powers. The word fetish became popular in Europe, where northern Protestants used it to criticize Roman Catholic practices such as that of the Corpus Christi (or Corpus Domini) feast, and in the eighteenth century, Charles de Brosses developed a theory of fetishism as a means to understand ancient religions. Marx was deeply influenced by the work of de Brosses for the conception of his idea of commodity fetishism, which he defined with the following words:

> A commodity appears, at first sight, a very trivial thing, and easily understood. Its analysis shows that it is, in reality, a very queer thing, abounding in metaphysical subtleties and theological niceties ... There it is a definite social relation between men, that assumes, in their eyes, the fantastic form of a relation between things. In order, therefore, to find an analogy, we must have recourse to the mist-enveloped regions of the religious world. In that world the productions of the human brain appear as independent beings endowed with life, and entering into relation both with one another and the human race.[39]

---

38 J. Baudrillard, *The Consumer Society: Myths and Structures*, London: Sage, 2016, p. 193.

39 K. Marx, *Capital*, vols I and II, London: Wordsworth Editions Limited, [1867] 2013, pp. 46–7.

Marx's words point to the relational but also mystical character of commodities, to their transcendental value, to the fact that the powers and values of human social relations are transferred to the fetish. This also illuminates the mechanism through which commodity fetishism becomes the capitalist surrogate for religious sacramentality, whereby the power of money replaces divine grace. As Roland Boer observes in his work on *Capital* and fetishism, for Marx, in its pure essence, the fetish can be reduced to nothing other than capital itself.[40] As Marx writes in the beginning of Chapter 24, Volume 3 of *Capital*:

> The relations of capital assume their most externalised and most fetish-like form in interest-bearing capital. We have here M – M', money creating more money, self-expanding value, without the process that effectuates these two extremes. In merchant's capital, M – C – M', there is at least the general form of the capitalistic movement, although it confines itself solely to the sphere of circulation, so that profit appears merely as profit derived from alienation; but it is at least seen to be the product of a social relation, not the product of a mere thing.[41]

While fetishism, as discussed earlier, involved the transfer of social powers of human relations to objects, we now have objects – be they real or illusory, material or imagined – establishing a relation with other objects, and transcending relations among humans. Money produces more money, and capital produces profit. This is, as Boer observes, the full realization of this transfer of powers: 'the complete abasement of human relations, so much so that those relations

---

40 R. Boer, 'Kapitalfetisch: "The Religion of Everyday Life"', *International Critical Thought* 1: 4, (2011), pp. 416–26.
41 K. Marx, *Capital: A Critique of Political Economy*, vol. III, in *Marx/Engels Collected Works*, vol. 37, London: Lawrence & Wishart, [1894] 1998, p. 391.

simply disappear from the scene'.[42] The human is no longer present or productive in the formula, human powers have been metaphorically and mysteriously transferred to the gods, and in this essentially pure and distilled form of the fetish, capitalism becomes a religion. Rather than being its consequence, capital (and its various manifestations as discussed above) becomes the social productive force of human labour, and it is in these terms that capitalism becomes the 'religion of everyday life'. Thus, the producer and the user of a commodity are not necessarily engaging in a full relation with one another when selling or buying something. In a similar way, and as I will discuss in more detail in Chapters 2 and 3, a donation to a charity can be interpreted as a fetish, as a means through which we establish a relationship with a distant place without doing so fully, disavowing the relations that have established the need for that same donation.

Capitalism thus emerges as a religion of modernity, one that is underpinned by guilt and the transformation of the Christian spirit into the spirit of capitalism, the cult of embodied value articulated in commodity fetishism, unlimited faith in the market, progress and technical solutions, and the appropriation of the notion of sacrifice. These are some of the key themes that I will develop in the next three chapters, and that will guide my critique of three interconnected case studies – the water charities Water.org and WaterAid, celebrity-led social network campaigns aimed at solving the 'global water crisis' and the sustainability policies of one of the largest bottled water companies, Nestlé Waters. In doing so, I will delineate the contours of the transnational networks of global water governance, while also highlighting the contradictions and absurdities encapsulated in these late capitalist projects.

---

42 Boer, 'Kapitalfetisch', p. 423.

Thirsty capital

I have argued here that there is not a 'global water crisis', but rather numerous structural water crises that affect the most vulnerable in both wealthy and poor countries across the globe. In spite of the sense of urgency about the 'global water crisis' that has entered the political and corporate discourse, we might very well argue that capital, as a process, thrives on water crises.

We are now well aware that water is not just a natural resource, but is deeply embedded in social, political and economic processes. While the former view has championed technocratic approaches to water management that focus on greater efficiency and rational use, the latter are illustrated by deepening processes of appropriation of water resources by powerful actors and the parallel dispossession of weaker or marginalized social groups. In 2019, a book titled *The Water Paradox* was published. The paradox in question, according to its author, Ed Barbier, is the following:

> With overwhelming scientific evidence pointing to growing overuse and scarcity of freshwater, why did the world not mobilize its vast wealth, ingenuity and institutions to avert this crisis? ... The global water crisis is predominantly a crisis of inadequate and poor water management.[43]

The book was reviewed widely, and the water politics scholar François Molle noted: 'The political dimension of water management is almost totally absent here, as if irrelevant and economic orthodoxy could prevail alone on its own virtue,' concluding that the book 'could have been written 25 years ago'.[44] Indeed, I might add, the paradox is precisely

---

43 E. Barbier, *The Water Paradox: Overcoming the Global Crisis in Water Management*, New Haven: Yale University Press, 2019, p. ix.

44 F. Molle, 'Review of *The Water Paradox: Overcoming the Global Crisis in Water Management* (Barbier, 2019)', *Water Alternatives*, item 77 (2019).

the opposite of that exposed by Barbier: throughout this chapter, it was clear that 'the world' is mobilizing its wealth, ingenuity and institutions to solve the 'global water crisis', and we have arguably never been so exposed to this issue. Many private companies are actually profiting from the idea of the 'global water crisis'. Citizens and concerned individuals are called to take action, and as we will see in Chapter 3, many have responded. So why are we not solving this problem? The short answer is that the solution does not lie in the amount of money that we mobilize, but rather on how and to what ends we mobilize it. The problem is political, and if we do not address it radically, we will never solve it.

2

Care: Good Samaritans and the Politics of Care

The emergence of global water charities

What does it mean to care – for others and for ourselves – in the 2020s? What is the meaning of care in an era of environmental collapse, when our 'planet is on fire', and why does care matter? 'Take care', 'be careful', 'I don't care!', 'I care for', 'you are careless': we are surrounded by care, we are constantly reminded about the importance of caring, and in any given day we engage – sometimes unconsciously – with several forms of caring. Care, as María Puig de la Bellacasa suggests, is simultaneously 'a vital affective state, an ethical obligation and a practical labour'.[1] This suggests that care can be driven by affection but also by a relational obligation, and in both cases there are things that we have to *do* and actions that we have to *perform* to engage in the act of caring. In other words, sometimes we care because we want to – and this can be prompted by situated forms of knowledge and love – and other times we care because we know that others expect us to do so. This latter type of care is normative and is indeed commonplace in what de la Bellacasa refers to as 'everyday moralizations' of the Global North. More often than not, care loses its transformative

---

1 M. Rusca and K. Schwartz, 'The Paradox of Cost Recovery in Heterogeneous Municipal Water Supply Systems: Ensuring Inclusiveness or Exacerbating Inequalities?', *Habitat International*, 73 (2018), pp. 101–8.

and emancipatory value which could potentially lead to alternative forms of organizing and ends up being a way to reproduce broader societal and cultural norms. Take, for example, a company's pledge to use only recycled plastics in their packaging, aimed at showing that they care, while I, as a consumer, might decide to buy only packaging made from recycled plastics to show that I care. The current socio-ecological crisis only serves to reinforce the assumption that we care for the planet if we also care about ourselves and our lifestyle, our wellness and our fitness. As a consequence:

> Those considered as traditional carers – women generally – or as typical professional carers – nurses and other marginalized unpaid or lowpaid care workers – are constantly moralized for not caring enough, or not caring 'anymore', or for having 'lost' some 'natural' capacity to care.[2]

Under neoliberalism, as Emma Dowling remarks, care and profitability are played off against each other.[3] This kind of 'iCare capitalism', as it was called by Michael Goodman, sets up a biopolitics of economic choice whereby consumers choose the 'care-full', affective products to buy and support various causes, while at the same time charities, celebrities, NGOs and corporations select the on-the-ground humanitarian and environmental issues they wish to confront and rectify.[4]

Thus, there are various types of caring, and these can have different material consequences. In mainstream understandings, care has been reduced to a neoliberalized performance

---

2 M. P. de La Bellacasa, *Matters of Care: Speculative Ethics in More Than Human Worlds*, Minneapolis: University of Minnesota Press, 2017, p. 9.

3 E. Dowling, *The Care Crisis: What Caused It and How Can We End It?*, London: Verso, 2022.

4 M. K. Goodman, 'iCare Capitalism? The Biopolitics of Choice in a Neoliberal Economy of Hope', *International Political Sociology*: 7, 1 (2013), pp. 103–5.

that needs to be validated, quantified and – if possible – measured, ideally in a monetary value. Among them, philanthrocapitalism – a term that denotes the need for philanthropy to become efficient, profitable, market-driven and investment-based – has emerged as one of the latest, and arguably liveliest, expressions of neoliberalism and cultural capitalism. The number of super-rich people – those whose assets are worth over US$30 million – grew steadily during the last decade, even during the Covid-19 pandemic. The world's 2,153 billionaires – most of whom live in North America and Europe – have more wealth than the 4.6 billion people who make up 60 per cent of the planet's population.[5] Philanthropy is thriving, and giving away money, as the *Economist* put it, continues to be fashionable among the rich and famous.[6] New ways of being charitable have been largely assimilated by non-profit organizations, and are now enmeshed into the broader logics that inform the third sector.[7] Grounded in the assumption that it is the market that sets the rules of the game, the ethos of philanthrocapitalism is that if you want to save the world, you must make money while you are doing it.

Fitting well with the white saviour trope, philanthrocapitalists increasingly influence the aid agenda and at the same time are not accountable to voters or shareholders, nor can they be caught in the nets of state bureaucracy. They can thus think 'big', quickly, 'revolutionarily' and, most importantly, in intellectual isolation. However, these ventures raise questions and require critical scrutiny. Everyone can

---

5 Oxfam, *Time to Care: Unpaid and Underpaid Care Work and The Global Inequality Crisis*, briefing paper, January 2020.

6 'The Birth of Philanthrocapitalism', *Economist*, 23 February 2006.

7 A. C. Budabin and L. A. Richey, *Batman Saves the Congo: How Celebrities Disrupt the Politics of Development*, Minneapolis: University of Minnesota Press, 2021.

have an informed opinion, but if you are 'a rich guy with an opinion', as Bill Gates referred to himself, then your opinion is likely to receive considerable attention.[8] Gates, who in his recent book *How to Avoid a Climate Disaster* explained: 'I own big houses and fly in private planes – in fact, I took one to Paris for the climate conference – so who am I to lecture anyone on the environment?', is a telling example of the dialectics of philanthrocapitalism. Building on this line of inquiry, Japhy Wilson conceptualizes philanthrocapitalism as an ideology that 'mobilises a disavowed enjoyment of global inequality'.[9] At the same time, philanthrocapitalism reduces transparency and participation, thus enforcing the rules of business onto the public domain, as in the case of agricultural policies in Sub-Saharan Africa and South Asia.[10] Furthermore, as Stasja Koot and Robert Fletcher have outlined, even when philanthrocapitalism superficially seems to steer towards genuine inclusion of disempowered communities – what they call 'popular philanthrocapitalism' – such initiatives 'allow for neoliberal capitalism to further extend its reach under the pretence of empowering those whom it marginalizes'.[11]

Global water governance is not exempt from this trend, and in recent years a new and heterogeneous range of actors has been attempting to solve the so-called 'global water crisis' through philanthropic initiatives. Global water charities in

---

8 B. Gates, *How to Avoid a Climate Disaster: The Solutions We Have and the Breakthroughs We Need*, New York: Knopf, 2008, p. 8

9 J. Wilson, *Fantasy Machine: Philanthrocapitalism as an Ideological Formation*, Third World Quarterly 35: 7 (2014), p. 1144.

10 A. Shaw and K. Wilson, 'The Bill and Melinda Gates Foundation and the Necro-Populationism of "Climate-Smart Agriculture"', *Gender, Place and Culture* 27: 3 (2020), pp. 370–93.

11 S. Koot and R. Fletcher, 'Popular Philanthrocapitalism? The Potential and Pitfalls of Online Empowerment in "Free" Nature 2.0 Initiatives', *Environmental Communication* 14: 3 (2020), p. 288.

particular have emerged as influential NGOs, foregrounding a neoliberal vision of charity and philanthropy, and supporting market-based solutions to improve access to water. These organizations can be led by philanthrocapitalists directly, but they can also be founded and governed by common people inspired by the philanthrocapitalist logic.

In this chapter I provide an overview of how these water charities operate and how they came to be so popular; in Chapter 3 I delve into their campaigns, examining the growing influence of celebrities on global water charities and governance. I will thus build on the discussion of global water governance laid out in the previous chapter to explore and understand how global water charities reproduce, and are produced by, the ideas and processes that have more generally determined the neoliberalization of the water crisis. I then outline the multiple actors, ideas and practices that operate in the global quest to solve the water crisis through philanthropic and socially responsible initiatives, often conducted in partnership with bottled water companies and international organizations.

To structure my argument, I weave three related concepts – care, sacrifice and redemption – into my discussion of contemporary water governance, highlighting the capitalist appropriation and domestication of these concepts in the name of growth, progress, and 'good deeds' for others. The partial hollowing out of the state from the delivery of basic water services opened the doors of water governance to private actors and markets, but also to international NGOs and charities. Indeed, during the last two decades several large water charities started to populate the crowded landscape of global water governance, growing in number and becoming increasingly supported through private and public funding. Charities, in general, are booming. In the UK, for example, between 2009 and 2015 the number of charitable organizations grew by 30 per cent, and their expenditure by 45 per cent.[12] In the first half of 2020 a total of £5.4 billion was donated to charity in the UK[13] and estimates suggest that US$471.44 billion were donated to US charities in 2020.[14] These organizations are portrayed as a force of transformative or 'alternative' development, based on people-centred and participatory approaches, promotion of grassroot-driven development and innovation in and commitment to basic service delivery' to the most vulnerable.

And yet, not everyone sees the 'added value' of NGOs or trusts their ability to promote alternative forms of development. Maria Rusca and Klaas Schwartz, for instance, argued that NGOs' dependence on international donors makes them vulnerable to their requirements, which are often prioritized over the demands and needs of the beneficiary community.[15] Furthermore, as they comply with donors' requirements, the development models promoted by NGOs tend to become more conventional and increasingly less effective in promoting alternative forms of development. On paper, NGOs are guided by two distinct and competing ideological perspectives. As Esteban Castro have explained with regard to water governance, one set of NGOs is concerned with empowering citizens and deepening democracy, and they see inclusive governance as a way to promote more equitable development.[16] Another set tends to emphasize

---

12 N. Banks and D. Brockington, 'Growth and Change in Britain's Development NGO Sector (2009–2015)', *Development in Practice* 30: 6 (2020), pp. 706–21.

13 Charities Aid Foundation, *UK Giving Report 2021*.

14 *Giving USA*, press release, 15 June 2021.

15 M. Rusca and K. Schwartz, 'Divergent Sources of Legitimacy: A Case Study of International NGOs in the Water Services Sector in Lilongwe and Maputo', *Journal of Southern African Studies* 38: 3 (2012), pp. 681–97.

16 J. E. Castro, 'Poverty and Citizenship: Sociological Perspectives on Water Services and Public-Private Participation', *Geoforum* 38: 5 (2007), pp. 756–71.

full-cost recovery of water services and the deregulation of private actors, through an approach that ultimately leads to the commodification of nature. Although these ideological perspectives are, in principle, opposed, the neoliberal and the empowerment agendas have – more or less explicitly – converged in promoting market-based governance.

Research on water charities is limited, but they are proliferating. Running a search for 'water' on Charity Navigator – a website that evaluates charitable organizations based in the United States – yields nearly 8,000 results. A similar search on Charities Aid Foundation – a charity whose aim is to bring together companies, private philanthropists, other charities, governments and not-for-profit enterprises – results in more than 1,400 resources. Furthermore, there are many NGOs that operate in the water sector even though they are not formally listed as 'water charities', and they receive a considerable amount of international funding. For instance, the volume of OECD Development Assistance Committee funding channelled through NGOs in the water supply and sanitation sector increased from US$308 million in 2010 to $338 million in 2016.[17] Overall, water charities share a commitment to solve the water crisis, yet they vary in scope, size and geographical focus. 'Global water charities' typically have their headquarters in the US and Europe, and they raise funds globally to finance and implement projects aimed at supplying water to underserved communities in urban and rural areas in the Global South.[18] Some of these charities have religious ties, while others are more closely related to corporate funding or rely on both. Water charities also operate in different ways, ranging from funding

---

17 OECD-DAC, *Aid for Civil Society Organisations: Statistics based on DAC Members' Reporting to the Creditor Reporting System Database (CRS), 2016–2017*, 2019.

18 K. Bunds, *Sport, Politics and the Charity Industry: Running for Water*, London: Routledge, 2017.

infrastructural projects (for example, a standpipe connected to a mechanized borehole) to cooperation with governments and other financial institutions through public–private partnerships for funding and operating small-scale decentralized water supply systems. Often these interventions are accompanied by training and educational initiatives.

In this chapter, I focus on two of the most prominent global water charities, Water.org and WaterAid, to outline the main elements of the political economy of care and sacrifice that have been mobilized to solve the water crisis. Both charities operate in the water sector at global and local scales, and they have connections at the highest political levels. Their funders and members are routinely invited participants to international water and economic fora such as the WoWF, the Stockholm Water Week and the World Economic Forum (WEF), but also to high-level governmental initiatives. They are involved in national and regional expert panels and working groups to advise policy-making in community level initiatives, including the actual implementation and management of water-related projects. Both charities are financed by private organizations and well-known philanthropists, and they receive public funding through international donor organizations. Both are exemplary of the increased neoliberalization of water governance, but the two are also quite different and reflect particular trajectories. Where the two charities differ the most is perhaps in their initial thrust. On one hand, Water.org originated through, and places emphasis on, individual initiative. On the other hand, WaterAid was founded by, and speaks in the name of, the British water industry. The two charities exemplify different types of caring: individually led (Water.org), and seemingly collective (WaterAid). Neoliberal logics were not at the forefront of WaterAid's activities when it was funded in the 1980s but became more present over time due to increased pressure to comply with donors' expectations.

Similarly, Water.org originated from a seemingly disinterested initiative but is now fully entrenched with market logics. Both start with a story.

Dirty water, Good Samaritans

Matt Damon often shares an inspiring anecdote to evoke his turning point as an environmental activist, even though the details surrounding this event have slightly changed over time. The usual narrative, up until the publication of the book *The Worth of Water* (co-authored with the CEO and co-founder of Water.org, Gary White) in 2022, had Damon filming a movie in Zambia, Southern Africa, in 2006. As Water.org's page on Damon explains:

> After multiple trips around the world, Matt witnessed what life was like for a community living in the global water crisis. While filming a movie in Sub-Saharan Africa, Matt spent time with families in a Zambian village. They lacked access to water and toilets. Matt's exposure to their daily lives inspired a commitment to helping solve the global water crisis.[19]

Damon, whose activism has developed in parallel to his film career, embodies the Good Samaritan: a man who has travelled and who does not pass by when he sees people who suffer, but instead cares for them, and does this with the help of others.[20] As Damon himself explains:

> I think what resonates with me most is when you see people living without clean water and they are forced to scavenge for

---

19 G. White and M. Damon, *The Worth of Water: Our Story of Chasing Solutions to the World's Greatest Challenge*, New York: Portfolio, 2022.

20 J. Gulam, 'Save the World with Ben and Matt: Ben Affleck, Matt Damon, and the Importance of Film Texts to Critical Discussions of Star Campaigning', *Celebrity Studies* 10: 4 (2019), pp. 543–58.

water and basically use up all of their time just doing and just trying to basically survive to the next day. You realize that they are in such a crippling cycle of poverty that it's just a death spin that they can't possibly get out of.[21]

Damon is not, however, the first celebrity to play the role of the Good Samaritan. Such a role had already been claimed by the 'original' rockstar and frontman, Paul Hewson, aka Bono.[22] As Paul Seales has shown in the book *Religion Around Bono*, during the last three decades Bono has developed strong relationships with powerful political and economic figures, including George W. Bush, Jeffrey Sachs, Jesse Helms and Bill Clinton, himself becoming an influential political figure that advocates for the neoliberal promise that free markets and for-profits will solve global poverty. The connection between Bono and Damon is made clear in White's and Damon's book *The Worth of Water*. In it, Damon explains that, rather than being in Zambia because he was shooting a movie, he was there following Bono's invitation:

> I was in Zambia because Bono – the rock star who spends his spare time fighting to end extreme poverty – had been pestering me to go ... Bono believes that seeing poverty up close can change a person's priorities, can compel them to go out and do something about it. So he and his colleagues at the organization he started, DATA – which would eventually become the ONE Campaign – had been pressuring me to join them on a trip to Africa.[23]

---

21 Water.org, *Full Interview with Matt Damon About the Water Crisis* – 2010.

22 C. E. Seales, *Religion Around Bono: Evangelical Enchantment and Neoliberal Capitalism*, Philadelphia: Penn State University Press, 2019. On the subject, refer also to H. Browne, *The Frontman: Bono (In the Name of Power)*, London: Verso, 2013.

23 White and Damon, *The Worth of Water*, pp. 2–3.

During his two-week trip, Damon and his group followed a tight schedule, and visited several slums and rural villages across Southern Africa. On one of their last days, the group learnt about the 'water issue' (lack of access to clean and safe water) and met with a family who lived in a remote Zambian village. Their fourteen-year-old daughter spent an hour every day walking to the well to fetch water for her family. Damon and the rest of the delegation accompanied her to the well, and once there, the actor tried to operate it:

> I had just finished filming one of the Jason Bourne movies, so I thought I was in pretty good shape. But pumping water from this well was harder than it looked. The girl and I laughed as I struggled with it.[24]

For Damon, this encounter was an eye-opener. He realized that water was central to everything and that life was impossible without it. As a celebrity, he felt he had an opportunity to redirect some of his 'needless surplus of attention' towards solving the water crisis and making a difference. Thus, a year later, when Damon was the executive producer of the documentary *Running the Sahara*, its locations (Senegal, Mauritania, Mali, Niger, Libya and Egypt) appeared to him as the 'Ground Zero in the global water crisis', and the project offered a good opportunity to start a fundraiser with his newly established foundation, H2O Africa.[25] Fast forward to 2022, and Matt Damon is an internationally recognized advocate for clean water worldwide: an outspoken 'influencer'.

In his book, Damon explains that he is a big admirer of the economist Jeffrey Sachs, one of his mentors in international development. Bono, who apparently calls himself a 'Jeff Sachs groupie', called Sachs 'the squeaky wheel that roars', while the entrepreneur and philanthropist Bill Gates aptly closed the circle, writing that 'Sachs is the Bono of

---

24 Ibid., p. 4.
25 Ibid., p. 21.

economics – a guy with impressive intelligence, passion, and powers of persuasion who is devoting his gifts to speaking up for the poorest people on the planet.'[26] But, as Japhy Wilson has noted, appearances can be deceiving. While Sachs poses as a saviour of the Global South – thus moving on from his role as a macroeconomist who delivered neoliberal economic shock therapy to countries in the former Soviet Union and South America – his recent market-based efforts aimed at ending global poverty in Africa exposed these countries to further economic exploitation.[27] In the book *The Idealist*, journalist Nina Munk provides a detailed account of how Sachs's Millennium Villages Projects – a US$120 million demonstration project aimed at providing a blueprint strategy to lift some of the poorest places on Earth out of poverty – came up short. Sachs, who often acted in competition with other development agencies and international organizations that were operating on the continent, tried to 'donate' people out poverty but underestimated the structural and contextual conditions that set these places in a state of poverty.[28]

Yet there is, undoubtedly, much to commend in Sachs's original good intention, or, for that matter, in anyone that decides to make a donation to a charity or to altruistically help and practise a monetized solidarity to those in need. As Gates put it, Sachs 'could have a good life doing nothing more than teaching two classes a semester and pumping out armchair advice in academic journals. But that's not his style. He rolls up his sleeves. He puts his theories into action.'[29]

---

26 B. Gates, 'A Cautionary Tale from Africa', gatesnotes.com, 21 May 2014.

27 J. Wilson, *Jeffrey Sachs: The Strange Case of Dr. Shock and Mr. Aid*, London: Verso, 2014.

28 N. Munk, *The Idealist: Jeffrey Sachs and the Quest to End Poverty*, New York: Anchor, 2013.

29 Gates, *A Cautionary Tale*.

I think about this often. Sachs, Damon, Bono or Gates are all doing far more than I have ever done to help those in need. They could have just minded their own business and nobody would have had a problem with that; they could still have had a good life. So why are they doing what they are doing? Why has Matt Damon decided to solve the water crisis? More generally, what drives philanthropic capitalism?

If we assume that capitalism has indeed become the religion of everyday life, and accept that its processual component, neoliberalism, is always historically and geographically situated and context-specific, the current proliferation of high-profile do-gooders is one of the ways in which the capitalist religion – and its accompanying cultural-material processes and representations – deals with the pressing humanitarian, environmental and development challenges that humanity faces in the Anthropocene. This is not necessarily due to rising disenchantment, but rather because of a 'migration of the holy', the virtuous and the pure 'good', to the domains of production and consumption, capital and markets, growth and crisis, and profit and price.[30]

Consider, for example, St Augustine's doctrine of the two cities: Rome (or the new Babylon), and Jerusalem (the city of heaven), which respectively symbolize all that is worldly, and all that is heavenly/Christian.[31] Under his interpretation, the two cities have been formed by two loves: the love of self (Rome), and the love of God (Jerusalem). St Augustine draws a distinction between being selfish and being social, between praising God and praising material wealth, between being generous and being envious. It is thus not possible to praise God and pursue the accumulation of wealth; or at least the

---

30 E. McCarraher, 'The Enchantments of Mammon: Notes Toward a Theological History of Capitalism', in *Modern Theology* 21: 3 (2005), pp. 429–61.

31 Augustine, *The City of God against the Pagans*, ed. R. W. Dyson, Cambridge: Cambridge University Press, 1998.

accumulation of wealth, per se, cannot grant salvation or be interpreted as a sign of the love of God. But charity – the highest form of love – and the donation of material wealth, can unite humans to God, as through this virtue we show our love to others, and also to God. In other words, while 'it is easier for a camel to go through the eye of a needle than for a rich person to enter the kingdom of God', things might get easier for the rich person if some of their wealth went to charity before trying to go through the eye of the needle.[32] With the sixteenth-century Reformation, Martin Luther, and later Calvin, reinterpreted these ideas and proposed the doctrine of the two kingdoms, to clearly define the boundaries between state and church: the kingdom of law and the kingdom of grace. God rules both realms, but the two are separate and should not be confused. Humans living in the material/immanent kingdom of law can still achieve righteousness using their free will and reason and deciding how to dispose of their wealth.

And if, as Weber suggested,[33] the accumulation of individual wealth is a sign of virtue in Protestant theologies, we can understand how the two main capitalist models (at least in theological thought), the Protestant/Anglo-Saxon and the Catholic/Latin, bring with them a different understanding of philanthropy.[34] In the Catholic sphere, the two 'kingdoms' are deeply entrenched, and charity is often connected to the Church or to political organizations or cooperatives: you do 'good' – and perhaps help those in need – through your work. In the Anglo-Saxon Protestant model, the two realms are separated. You run your business in the kingdom of law and seek redemption in the kingdom of grace. Consider

---

32 Matthew 19:24, Bible.
33 M. Weber, *The Protestant Ethic and the 'Spirit' of Capitalism and Other Writings*, New York: Penguin, 2013.
34 L. Bruni, *L'arte della gratuità*, Vita e Pensiero, Milan, 2021.

the example of Jeff Bezos. On one hand Bezos, the CEO of Amazon, builds a fortune through his major plastic polluter company,[35] launches himself into space through his Blue Origin initiative,[36] and is involved in a series of controversies over the way in which his company crushes unions and the rights of his workers.[37] On the other hand, and after his fortune was made, Bezos, the man on a journey, allocates some of his wealth through charitable initiatives such as the Courage and Civility Award – a US$100 million award to individuals who tackle challenges facing the human race – or the Bezos Earth Fund, a US$10 billion initiative aimed at driving 'transformational system change', as 'this is the decisive decade' to fight climate change and protect nature.[38] But these two domains are only theoretically separated. In practice they are not, and in this case, they are united by a contradiction.

Jeff Bezos thrives in a system that allowed him to accumulate a personal wealth of over US$150 billion in less than three decades (Amazon was founded in 1994), a wealth that increased by more than 30 per cent in the first two years of the Covid-19 pandemic.[39] Bezos is one of the roughly 2,700 billionaires in the world. And this speaks to global inequalities – that today are as great as they were at the peak of Western imperialism in the early 1900s[40] – but also

---

35 Oceana, *Exposed: Amazon's Enormous and Rapidly Growing Plastic Problem*, December 2021.

36 S. Neumann, 'Jeff Bezos and Blue Origin Travel Deeper into Space Than Richard Branson', NPR, 20 July 2021.

37 D. Streitfeld, 'How Amazon Crushes Unions', *New York Times*, 16 March 2021.

38 B. Luscombe, 'What Jeff Bezos' Philanthropy Tells Us About His New Priorities – and What Change They May Bring', *Time*, 27 July 2021.

39 C. Collins, 'Updates: Billionaire Wealth, U.S. Job Losses and Pandemic Profiteers', inequality.org, 21 November 2022.

40 'Executive Summary', *World Inequality Report*, 2022.

to the climate crisis, as the wealthiest 1 per cent emit more than twice as much CO_2 as the bottom 50 per cent.[41] More to the point, while economically efficient and logistically excellent, the consumerist logic behind Amazon's business model – buy more, pay less, buy more often – is arguably one of the main causes of environmental degradation.[42] And yet, Bezos purportedly wants to change the (outcomes of the) same system that made him rich, and with the Bezos Earth Fund he made the biggest ever individual donation to transform how humanity attempts to address the climate crisis. But the climate crisis is first and foremost a political issue.[43] Dealing with it requires a full reorganization of our social metabolism, and this can only be achieved through collective action rather than at the initiative of an individual.

But let us return to Damon and Water.org. In 2022, the publication of *The Worth of Water* put the charity in the spotlight. The former president of the United States, Bill Clinton, for instance, welcomed the book enthusiastically, and invited Damon and White to his podcast 'Why Am I Telling You This?'. In his review of the book, Clinton wrote:

> I feel lucky to have been there for that first, fateful meeting – at the Clinton Global Initiative in 2008 – between a movie star with a passion for global development and a water and sanitation engineer with years of on-the-ground expertise. More than a decade later, that unlikely pair has helped transform the lives of millions of people around the world through safe water and sanitation. Matt and Gary's vision is clear, their strategy is smart

---

41 Oxfam, *Ten Richest Men Double Their Fortunes in Pandemic While Incomes of 99 Percent of Humanity Fall*, press release, 17 January 2022.

42 G. Kallis, 'In Defence of Degrowth', *Ecological Economics* 70: 5 (2011), pp. 873–80.

43 M. Huber, *Climate Change as Class War: Building Socialism on a Warming Planet*, London: Verso, 2022.

and effective, and their faith in their fellow human beings is at the heart of everything they do.[44]

There is almost a messianic undertone in the narrative surrounding the birth of Water.org, starting from Damon's calling while in Zambia and continuing with the seemingly inevitable encounter between Damon and White, and the cascade of events that it triggered. It thus seems natural for Damon to hold the moral high ground and to advocate for allegedly universal values and ideas. Speaking to Andy Serwer, the host of the Yahoo Finance show *Influencers*, Damon explained that the water crisis 'is not a political issue at all' and that 'the left doesn't have a monopoly on compassion': what Water.org does, with their market-driven solutions, system and programming, 'is very effective and efficient', and therefore this should be widely accepted, as people believe in 'ideas that work'.[45] But as I have illustrated earlier, the water crisis has deep political roots and implications. While it is true that nobody has – or should have – a monopoly on compassion, the act of caring is essentially a political act. So, let us have a look at how Water.org cares for those in need of clean and safe water, exploring how their visions of care materialize on the ground, and what relational configurations they constitute in the process of caring.

I care: Water.org

Water.org is a global non-profit organization whose aim is to bring clean water and sanitation to the world. The charity was founded by Damon and Gary White, a civil and environmental engineer, with experience in water and sanitation

---

44 White and Damon, *The Worth of Water*, back cover.
45 M. Zahn, 'Matt Damon: The Political Left "Doesn't Have a Monopoly on Compassion"', *Yahoo! Finance*, 7 April 2022.

projects. While Water.org was founded in 2009, the charity is the result of the merger of WaterPartners International, a charity founded by White in 1990, with H2O Africa Foundation, an NGO founded by Damon in 2006. Thus, Water.org functions through White's entrepreneurial expertise in microfinance in the water sector, which is amplified and made fully global thanks to Damon's celebrity status and commitment to this cause outlined in the previous section. Damon's original good intention – inspired by what appears to be genuine compassion – is not only admirable,[46] but also reveals the engulfing and parasitic nature of capitalism, and its capacity to feed on other organisms and ideas.[47] This is because Damon's call to action to solve the water crisis, with its philanthropic logic, has very rapidly been absorbed by capitalism and its market logic. As Damon himself has put it, 'The problems are so massive in dollar terms, you will never get there with philanthropy alone. You have to bring in the real money.'[48] Through similar logic, as Gary White explains on Water.org's website, 'People living in poverty are not a problem to be solved. They are a market to be served.' Caring is synonymous to making a profit.

And indeed, the aim of Water.org is to use market forces to solve the water crisis. Those who suffer from water scarcity and poor access to sanitation are seen as a business opportunity, rather than people to be helped. Water.org operates

---

46 Colleagues consider Damon an example. The actor Christian Bale describes his philanthropic involvement thus: 'He's also one of the bloody most decent men I've ever come across … He's so cooperative. He's absolutely in the school of you cooperate with each other, you don't compete against each other at all … The way he views everything, he has this bigger perspective' (L. Morrison, 'Water Relief from a Star – Profile: Matt Damon', *Lifestyles Magazine* 285 (2020), p. 52).

47 Z. Baumann, 'Capitalism Has Learned to Create Host Organisms', *Guardian*, 11 October 2011.

48 B. Booth, 'Matt Damon's Water Crusade Has Helped 16 Million People So Far', CNBC, 24 January 2019.

through grants and, predominantly, through its WaterCredit Initiative (WCI), a market-driven solution that 'empowers people to immediately address their own water needs'.[49] Water.org identifies regions where people need access to water and sanitation, and partners with local institutions (generally banks) to finance small projects such as wells, toilets and related infrastructure. When a loan (the average loan size is US$365) is repaid, the money is lent to another borrower, then to another, and so on. The process is very straightforward and has three main implications: i) those who receive loans see an immediate and tangible benefit in their everyday life; ii) the lending institutions make a considerable profit; and iii) the growth of this mechanism is exponential, as is the reach of Water.org. While in 2009, Water.org served 137,400 people with clean water and sanitation,[50] the cumulative total number of people served reached 5 million in 2016,[51] 10 million in 2017,[52] and 17 million in 2018.[53] Likewise, its total revenue and expenses have grown considerably year after year.

Water – a chemical substance and a social resource – is thus transformed into a 'smart investment', in a move that seems particularly beneficial for the banks that lend the money.[54] As Damon puts it:

> We underwrote a lot of these loans and worked with local partners in these communities and they were so successful that now we've gotten out of the way and commercial capital has come in and you know, I sat with a branch manager of a bank in India who said I'm gonna call every branch manager in India and let

---

49 Water.org, *The WaterCredit Initiative*, 2009.
50 Water.org, *Annual Report 2009*.
51 Water.org, *Annual Report 2016*.
52 Water.org, *Annual Report 2017*.
53 Water.org, *Annual Report 2018*.
54 F. Menga and E. Swyngedouw (eds), *Water, Technology and the Nation-State*, London: Routledge, 2018.

them know that these are really great loans because they pay back at such a high rate and you're being introduced to a whole new level of customer.⁵⁵

Credit – or in this case microcredit – emerges as the fetish, as the social productive force of human activities and labour, whereby banks operate as the churches led by celebrity high priests of neoliberal capitalism (such as Damon) and their 'lesser' priests (such as White), who administer the sacrament of credit debt. Money lent through microcredit produces more money and a profit for the lending institution, thus expanding the reach of capitalism into new and untapped markets. To further press this point, a Water.org staff member published a research article in *Aquatic Procedia* arguing that microloans in the water sector are less risky than commonly perceived, and water is, as such, investment-worthy.⁵⁶

While microfinance's rhetoric of empowerment and its validity as a universal development tool have been questioned by critics both as a practice and an ideology, in its purest, distilled form, microfinance becomes an almost transcendental being, an abstract and universal entity that has the power to establish relationships between peoples and places and to solve the 'global water crisis'.⁵⁷ In other words, the fetishization of microfinance leads to a simplification, and to an extent a disavowal, of the complex relations behind a specific microcredit project and the 'global water

---

55 Water.org, 'Full Interview with Matt Damon About the Water Crisis – 2010'.

56 L. Pories, 'Income-enabling, not Consumptive: Association of Household Socioeconomic Conditions with Safe Water and Sanitation', *Aquatic Procedia* 6 (2016), pp. 74–86.

57 N. Aslanbeigui, G. Oakes and N. Uddin, 'Assessing Microcredit in Bangladesh: A Critique of the Concept of Empowerment', *Review of Political Economy* 22: 2 (2010), pp. 181–204; A. Schwittay, *New Media and International Development: Representation and Affect in Microfinance*, London: Routledge, 2014.

crisis' in general. The limit of this approach is that it attempts to solve the water crisis by treating its symptoms (lack of safe water and sanitation), rather than questioning its causes: poor governance, rapid urbanization, privatization, water grabbing, reckless consumption and economic and political inequality.[58] This contradiction is eloquently illustrated by the WaterEquity Global Access Fund, a US$150 million private investment fund launched by Water.org in 2019 to:

> provide debt capital to high-performing financial institutions in emerging markets to enable them to scale their water and sanitation micro-finance portfolios. The size of this undercapitalized market is large, with an estimated USD18 billion of demand from families living in poverty.[59]

Water.org is committed to 'accelerating an end to the global water crisis for millions of women, children, and men', and it is aiming to do so through a seemingly win-win scenario: pushing forward the capitalist frontier and targeting countries with large populations living without access to safe water and sanitation.[60] While solving the water crisis is still prominent in Water.org's mission (the original good intention discussed above), this seems to have become a means – to expansion into new large and untapped markets – rather than an end.

Among its investors, WaterEquity includes the Bank of America, the Conrad N. Hilton Foundation and Niagara

---

58 K. Bakker, *Privatizing Water: Governance Failure and the World's Urban Water Crisis*, Ithaca: Cornell University Press, 2010; R. Boelens et al., 'Hydrosocial Territories: A Political Ecology Perspective', *Water International* 41: 1 (2016), pp. 1–14.

59 WaterEquity, *US Asset Manager WaterEquity Launches USD 150 Million Impact Investment Fund to Accelerate Access to Safe Water and Sanitation for All*, press release, 13 November 2019.

60 Target regions include Latin America and the Caribbean, sub-Saharan Africa, Middle East and North Africa, Eastern Europe and Central Asia, South Asia, and East Asia and the Pacific.

Bottling. The participation of Niagara Bottling (a manufacturer of bottled water and soft drinks) as a contributing partner and foundational investor is particularly telling of the abovementioned contradiction. As we shall see in Chapter 4, plastic bottles are third in the list of items most discarded and collected from the oceans, and they require large amounts of energy to be produced (and so generate greenhouse gas emissions). More generally, bottled water arguably leads to the commodification of water, making it pricier and more inaccessible, while constraining local water supplies across the globe.[61] And besides the general problems that we may have with bottled water, Niagara Bottling has been directly involved in several notable controversies, conflicts and even legal battles with local communities about water pumping in aquifers across the United States, including the Floridan Aquifer in Florida[62] and the Cooper Lake in Kingston, New York.[63] As Rebecca Martin, a Kingston resident and community activist who fought Niagara noted:

> To privatize a municipal water supply for a corporation to commoditize in plastic bottles and ship from our watershed to other parts of the world is an awful idea for everyone. It's a short-term solution that benefits capitalists and a real wake-up call for our community.[64]

---

61 H. Gleick, *Bottled and Sold: The Story behind our Obsession with Bottled Water*, Washington, DC: Island Press, 2010; D. Jaffee and R. A. Case, 'Draining us Dry: Scarcity Discourses in Contention over Bottled Water Extraction', *Local Environment* 23: 4 (2018), pp. 485–501.

62 K. Spear, 'Bottler Wins OK to Pump Twice as much from Aquifer', *Orlando Sentinel*, 12 February 2014.

63 A. Okeowo, 'A Jazz Singer Fights Niagara Bottling', *New Yorker*, 1 July 2016.

64 R. Martin, personal communication, 24 April 2020.

While Niagara's Director of Corporate Giving describes the partnership with WaterEquity as a means to accelerate 'an end to the global water crisis for hundreds of millions of women, children, and men [living in] emerging markets', Niagara's actions as a profit-seeking business seem to make water more exclusive and less equally distributed.[65]

The mutually constitutive relationship between the marketization of the water crisis and the commodification of water resources is held together by an acritical belief in the healing capacities of progress and technical solutions. A polyethylene terephthalate (PET) water bottle, for example, is not conceived of as an unnecessary source of plastic pollution, but rather as an innovation that provides the 'smallest ratio of packaging material to product'. According to Niagara, a PET bottle comprises 2 per cent package and 98 per cent product (as compared to an egg which it claims comprises 13 per cent package and 87 per cent product!).[66] In a related vein, the installation of taps, toilets and the network of pipes necessary for the circulation of water becomes both the material accomplishment of the desire to end the water crisis, and a market with an enormous untapped potential. As White and Damon highlight:

> Clean water and sanitation is a viable market, and we are not alone in understanding its potential. The Bill and Melinda Gates Foundation estimates that the global market for micro-finance in water is US$12 billion. Deloitte's Monitor Inclusive Markets estimates a US$10 billion to US$14 billion market for toilets in rural India alone. The demand exists, as does the approach to meet it. Now we are beginning to see the missing piece: the investment capital that will catalyze the process.[67]

---

65 WaterEquity, 'Investors'.
66 Niagara, 'Packaging Efficiency', Niagara's website.
67 White and Damon, *The Worth of Water*, p. 56.

Damon thus emerges not only as a celebrity post-humanitarian high priest, but also as an expert on water and markets, and this is further evidenced by Water.org campaigns, as I will illustrate in the next chapter. His foundational vocation – solving the global water crisis – is still central, but at the same time feels distant, almost peripheral, and spatially it is unfolding in a generic and utopian 'developing world' in need of help. H2O, it appears, has been channelled into a form of apolitical populist environmentalism and global developmentalism, whereby the ecological, economic and social problems caused by capitalism and modernity are detached from the relations of global neoliberal capitalism that lie behind them, and the matter of concern is disembodied and ultimately vague.[68]

We care: WaterAid

While Water.org is arguably the most visible global water charity today, there is another that has been around for more than four decades and paved the way for the current fortunes of the sector. To understand its genesis, we need to go back to 27 January 1981, when the Thirsty Third World (TTW) conference took place in London at the initiative of the National Water Council (NWC), an independent statutory body that brought together various representatives of the water industry in England and Wales in an attempt to bridge the gap between water authorities and the central government. In turn, the TTW conference was inspired by the 1977 UN Water Conference in Mar del Plata, which recommended that the period 1980 to 1990 should be

---

68 E. Swyngedouw, 'The Antinomies of the Postpolitical City: In Search of a Democratic Politics of Environmental Production', *International Journal of Urban and Regional Research* 33: 3 (2009), pp. 601–20.

designated the International Drinking Water Supply and Sanitation Decade, and 'should be devoted to implementing the national plans for drinking water and sanitation'.[69] The TTW conference was organized by David Kinnersley, an economist and public servant who previously headed the North West Water Authority, and who had been touched by how people in the developing world had to struggle to get clean and safe water – a story not dissimilar to Damon's own foundational myth. The report of the TTW conference is an interesting read, particularly because it provides a snapshot of the dominant thinking in early 1980s global water governance, which at the time was set upon the goal that the water crisis in the 'third world' had to be solved by 1990.[70]

Readers might be familiar with the forward-looking projections and the long-term targets that dominate how we talk about the climate crisis and environmental degradation. There are 'only 11 years left to prevent irreversible damage from climate change' noted, for instance, the UN General Assembly President María Fernanda Espinosa Garcés in 2019.[71] Ten years earlier, in March 2009, Charles, then Prince of Wales, warned that the world has only '96 months to act' (so until July 2017) before the damage caused by global warming becomes irreversible.[72] As early as the 1980s, analogous, even if less dramatic, warnings about the ephemerality of life on planet Earth and the need

---

69 UN Secretary-General, 'International Drinking Water Supply and Sanitation Decade: Present Situation and Prospects', *Report of the Secretary-General*, 1980, p. 5.

70 National Water Council, *Thirsty Third World: A Report of the NWC Conference Held in London in the 27 January 1981 to Support the Start of the Water Decade 1981–1990*, London, 1981.

71 UN, *Only 11 Years Left to Prevent Irreversible Damage from Climate Change, Speakers Warn during General Assembly High-Level Meeting*, press release, 28 March 2019.

72 R. Verkaik, 'Just 96 months to save world, says Prince Charles', *Independent*, 9 July 2009.

to take action were already in abundant circulation. My point here is not to deny the veracity – or even the necessity – of these many warnings; rather, it is to question their effectiveness.

But let us return to the TTW conference. In the report, Peter Bourne – at the time an assistant secretary-general at the United Nations – noted that the 'numbers of people without wholesome water or sanitation are such that every single day from now till 1990, improvements need to be brought to another half million of them'.[73] In case this was not going to happen, what mattered was that an 'irrevocable process' was set in motion 'so that whether it is five years beyond the [water] Decade or ten, it will be absolutely certain that complete coverage will be achieved'.[74] According to the report, the solution was clear and the technology existed. Even the money was there: 'Each day we spend $240 million on cigarettes, more than enough to pay for the Decade'.[75]

And yet, as we now know, the issue of clean water and sanitation is still far from being solved, but we will return to this point later. Among the many ambitious decisions that were taken or proposed at that meeting, one was to stir public opinion by exposing the scale of suffering from poor water and sanitation. Baroness Eirene White – a British Labour politician and journalist – in particular felt that more needed to be done to involve 'voluntary-aid' bodies and organizations, and encourage voluntary giving with well-run information campaigns and through the advice of organizations such as Oxfam and Save The Children. Was this the beginning of water charities as we know them today?

We will never know for sure, but what we know is that six months later, on 21 July 1981, the charity WaterAid was

---

73 National Water Council, *Thirsty Third World*, p. 4.
74 Ibid.
75 Ibid, p. 8.

officially established by members of the UK water industry, including Affinity Water, Portsmouth Water, SES Water and South East Water, and with the support of Sir Robert Marshall, chairman of the NWC. Its aim is to 'transform lives by improving access to safe water, hygiene and sanitation in the world's poorest communities'.[76] While, like Water.org, this venture also started with a Good Samaritan (Kinnersley), I suspect that most readers will have to dig deep to link the name of WaterAid to a specific individual (except, perhaps, for the many celebrities that support their campaigns). As a retired development worker observed: 'It's very hard to name a high-profile individual that works for them, except for the project officers that I've met during my work'.[77] WaterAid is indeed the result and expression of a collective initiative led by the UK water industry and its representatives that, in turn, were following the lead set by the UN Drinking Water Supply and Sanitation Decade. WaterAid's strategy and deeds thus provide a good example of the results that compassion and care in the water sector can trigger when they come directly from the 'system'. What makes WaterAid – and Water.org as well – a particularly compelling case for this book is that they are both a symptom and a cause of the current water crisis, and thus encapsulate some of its main contradictions.

Indeed, England and Wales are the only countries in Europe where the water sector has been entirely privatized. Water privatization took place in 1989, but its occurrence had been long in the making, as this was part of the British Prime Minister Margaret Thatcher's campaign against state ownership of strategic assets such as telecommunications and transport infrastructure. As Thatcher recalls in her memoirs:

---

76 'Modern Slavery Act statement', wateraid.org.
77 Interview with author, January 2021.

> Privatization ... was fundamental to improving Britain's economic performance. But for me it was also far more than that: it was one of the central means of reversing the corrosive and corrupting effects of socialism ... But through privatization ... the state's power is reduced and the power of the people enhanced. Just as nationalization was at the heart of the collectivist programme by which Labour Governments sought to remodel British society, so privatization is at the centre of any programme of reclaiming territory for freedom.[78]

More prosaically, the British government had been underinvesting in its hydraulic infrastructure since the 1970s, and river and tap water quality went through a steady decline. As geographer Karen Bakker explains, when the European Union introduced more stringent legislation on water quality in the mid-1980s, it had to open a noncompliance procedure against Britain, which should have invested between £24 to £30 billion (at 1989 prices) to meet European water quality standards.[79] This would have required massive public sector borrowing – something that the Thatcher government was determined to avoid. Privatizing the water sector thus offered a good opportunity to avoid a tax increase, while also allowing the government to 'reclaim' liquid 'territory for freedom'. As a result, water became more costly for consumers in England and Wales: on average, prices rose by over 50 per cent between 1989 and 1993, and by 46 per cent in real terms, adjusted for inflation, between 1989 and 1998.[80] But this was not matched by increased investment to maintain water pipes and sewers in a good state. On the contrary,

---

78 M. Thatcher, *Margaret Thatcher: The Downing Street Years*, New York: HarperCollins, 1993.

79 K. Bakker, 'Neoliberalizing Nature? Market Environmentalism in Water Supply in England and Wales', *Annals of the Association of American Geographers* 95: 3 (2005), pp. 542–65.

80 E. Lobina and D. Hall, 'UK Water Privatisation: A Briefing', Public Services International Research Unit (PSIRU), 2001.

some water companies cut their investment programmes to maintain or increase their dividends. As reported by the *Financial Times* in 2021:

> Britain's water and sewage companies have slashed investment in critical infrastructure by up to a fifth in the 30 years since they were privatised ... The decline was most extreme for wastewater and sewage networks. Investment there has fallen by almost a fifth, from £2.9bn a year in the 1990s to £2.4bn now ... The reductions have come despite a 31 per cent real-term increase in water bills since the 1990s – an average of £100 a year per household – and £72bn in dividend payments to parent companies and investors including private equity, sovereign wealth and pension funds in the same period.

Furthermore, recent research showed that the poor management of untreated wastewater and raw sewage by water companies is the main source of microplastic pollution in the UK's rivers; the water companies are thus, paradoxically, the cause of this contamination.[81] So, while water companies in England and Wales are keen to 'transform lives' and improve access to safe water, hygiene and sanitation in the world's poorest communities, they arguably have different priorities for their domestic market. Unsurprisingly for private companies, they strive to maximize profits, even if this means that prices go up in order to pay more dividends to their investors. But things are different when it comes to their charity WaterAid, whose income comes, in large part, from individuals, and is then redistributed to their partners in their target countries.[82] Profit and corporate giving are,

---

81 R. Hurley, J. Woodward and J. J. Rothwell, 'Microplastic Contamination of River Beds Significantly Reduced by Catchment-Wide Flooding', *Nature Geoscience* 11 (2018), pp. 251–7; University of Manchester, *Water Companies Are Main Cause of Microplastic Pollution in UK Rivers*, press release, 13 May 2021.

82 As stated in their latest available annual report, 47 per cent of

quite literally, two sides of the same coin. One of the first initiatives of WaterAid was to urge customers to round up their water bill to the nearest British pound in support of its campaigns, something that allowed it to raise £800,000 a year by 1997.[83]

Today, WaterAid has grown into a large and influential organization in global water governance, with projects in twenty-five countries and a consolidated annual budget of nearly £100 million.[84] Its mission is to focus on multiple dimensions of the water crisis (access, sanitation, climate resilience, public health and security) through a range of activities including service delivery, evidence-based policy, and campaigning work. Over the years, WaterAid gave membership to WaterAid America and Australia (2004), Sweden (2009), Japan (2013) and India (2016), thus becoming a truly global organization. Additionally, its key members, the UK water companies, are no longer *truly* British. Almost three quarters of England's water industry is currently owned by foreign companies. Just to cite a few, Wessex Water is owned by the Malaysian company YTL Corporation Berhad, while 80 per cent of Northumbrian Water is owned by CK Hutchison Holdings Limited – a Hong Kong–based and Cayman Islands–registered multinational corporation – and around half of Thames Water is owned by sovereign wealth funds based in the United Arab Emirates, Kuwait, China and Australia.[85] In 2021, three of the biggest water companies, Thames Water, Yorkshire Water and Southern Water, paid no corporation tax.[86] Thames Water, in particular – the UK's

---

WaterAid's income comes from individuals, 13 per cent from major donors and trusts, 15 per cent from corporate donors and 25 per cent from institutions and other sources.

  83 'David Kinnersley – Obituary', *Telegraph*, 17 December 2004.
  84 WaterAid UK, *Annual Report 2020–2021*.
  85 GMB, *More Than 70% of England's Water Industry Owned by Foreign Companies*, 17 September 2018.
  86 N. Craven, 'Revealed: Three British Water Firms That Pay NO

largest water and wastewater company – took advantage of the British government's capital allowances scheme and has paid zero corporation tax during the last decade or so. The company did pay, however, more than £6 million in fines in 2021 after untreated sewage polluted a park and rivers, killing fish and other water life.[87]

So, if the water crisis is also the result of longstanding exploitative relations, unequal power dynamics, uncontrolled capture of resources, privatization of public assets, profit-driven market logics and insufficient infrastructural investments, then how fitting that privatized British water companies are among those leading the quest to solve it. In this respect, what matters is not necessarily what WaterAid *does* or *does not* do, or to question their good intentions. Rather, what matters is the fact that the water crisis – which is both a localized and context-specific problem and a global matter of concern deeply rooted in postcolonial and globalized transnational markets – is a problem of capitalism and for capitalism. As such, the 'global water crisis' exposes the largely contradictory and often unequal dynamics of capitalism, while also revealing its engulfing and parasitic nature, together with its capacity to feed on other organisms and ideas. Charity, when taken seriously with meaningful sacrifices and donations, threatens the logic of growth and accumulation that underpins the same existence of capitalism. Capitalism thus treats charity, in its essence, as a new shoot that must be cut off from the rhizome. And yet, before doing this, capitalism absorbs this potentially revolutionary new sprout and establishes a market on it.

Tax – Our Investigation Shows Foreign Owned Thames, Yorkshire and Southern Avoid Business Levy', *This Is Money*, 30 January 2022.

87 Department for Environment, Food and Rural Affairs, The Water Services Regulation Authority and Rebecca Pow, *Progress Made but Too Many Water Companies Still Falling Short*, press release, 13 July 2021.

Caught in this system, water charities become entrenched in the market logics and mechanisms of care driven by the imperatives of growth and competition. The disinterested act of giving is thus neutralized; it becomes a business that helps keep the system in place. To further exemplify this, one can look at Corporate Social Responsibility (CSR): large multinational companies have embraced the principles of CSR to tokenistically respond to social and ecological criticisms and ultimately expand their businesses towards green capitalism.[88] In the case of water charities, their relationship with capitalism contributes to the transformation of the very nature of giving in the water sector.

There is indeed a tension between WaterAid's normative values – that place the world's poorest and most vulnerable at the centre of its work – and its accountability to donors, which, ultimately, are its 'clients'. As Maria Rusca and Klaas Schwartz observed in their study of WaterAid's activities in Malawi and Mozambique, the charity's reliance on donors and the pursuit of cognitive legitimacy (the desire to be acknowledged as an expert by the system) push WaterAid to prioritize capitalist logics and the development strategies that donors' organizations embody and promote, rather than fostering inclusive development and system change.[89] Water-Aid's strategy, for instance, focuses on technical-financial, behavioural and managerial changes, rather than systemic change.[90] Despite an intimate knowledge of the water sector, the critique of the system which has produced water and sanitation inequalities is very limited. The reports published by WaterAid frequently mention facts and statistics about water, toilets and hygiene inequalities, but fail to mention,

---

88 E. Chiapello, 'Capitalism and Its Criticisms', in P. Du Gay and G. Morgan (eds), *New Spirits of Capitalism? Crises, Justifications, and Dynamics*, London: Oxford University Press, 2013, pp. 60–81.

89 Rusca and Schwartz, 'Divergent Sources of Legitimacy'.

90 WaterAid UK, *Everyone, Everywhere 2030*, 2017.

or question, the fundamental processes that underpin the crisis. As stated in WaterAid's 2017–18 strategic report, it is '[b]y strengthening every part of the systems involved, from the national to the local and adapting to each context' that 'we make sure communities feel the benefits long after we have gone'.[91] Likewise, the report continues, 'relatively small, targeted investments into strengthening systems can substantially improve service efficiency, effectiveness, accountability and sustainability, and reach more people for longer, with less wastage'.[92] Further confirming these concerns, research conducted by Tariya Sarauta Yusuf, Anna Murray and Chukwumerije Okereke on WaterAid Nigeria shows that while WaterAid emphasizes the need for partnerships and participation between states and citizens, in practice its well-intentioned approach unintentionally produced new power relationships and inequalities that rendered some of its efforts ineffective, while also reproducing old practices of exclusion among local communities.[93]

Maria Rusca, Rossella Alba and I have argued elsewhere that these narratives tend to challenge a system that is 'performing' poorly, rather than the system itself.[94] In line with this, WaterAid's projects consistently promote capitalist logics, including for profit social enterprises, private entrepreneurship, and full cost recovery principles. Although WaterAid seeks to combine these capitalist logics and its mission/visions into a 'win-win reform framework' that works for people (via universal access) and its commercial

---

91 WaterAid UK, *Annual Report 2017–2018*.
92 Ibid., p. 11.
93 T. Sarauta Yusuf, A. Murray and C. Okereke, 'Working with Local Governments to Increase Access to wash Services: A Case of Wateraid's Participatory Approaches in Nigeria', *H2Open Journal* 5: 3 (2022), pp. 424–37.
94 F. Menga, M. Rusca and R. Alba, 'Philantrocapitalism and the Remaking of Global Water Charity', *Geoforum* 144 (2023), pp. 1–10; WaterAid UK, *Annual Report 2019–2020*.

partners (market expansion), financial objectives often prevail over social ones.[95] When WaterAid Malawi established community-based Water Users Associations (WUAs), these were tasked with selling water at the kiosks and collecting revenue for the water utility, which supplied kiosks with water at a subsidized price. While on paper these organizations might appear as a successful case of the commoning of water services, paradoxically, low-income dwellers ended up paying 4–5 time more for water than high-income residents as the cost of running the WUAs were passed on to water users. WUAs are run by local elites, motivated to increase water tariffs to their customers in order to augment their profit margins and salaries. The hierarchical leadership of WUAs and their profit motif thus lead to a local reterritorialization of water services into a commodity that undermines the ability of poor urban dwellers to access water.[96]

Conclusions

Care for other people and environments materializes not just in the bodies, media appearances and cultural capital of celebrities or CEOs, but also in the associated products, donations, goods, elites, consultants and corporations that circulate in a political economy of affect and concern that then touches down in the places 'most in need' of support, aid, capital and/or market-creation. Care materializes downstream in the 'things' of aid (such as water pumps, microloans and toilets) through the upstream divine powers of celebrity-led and corporate-approved advocacy of

---

95 WaterAid UK, *Reaching the Urban Poor: Supporting utilities to Engage Communities in Service Extension. An Overview of the WaterAid Modules for the World Bank Institute Course on Successful Utility Reform in Water Supply and Sanitation*, 2011.
96 Rusca and Schwartz, 'Divergent Sources of Legitimacy'.

markets and market-making and choice and choice-making, in a neoliberal (philanthropic and humanitarian) world. Put another way, the political economy of care created by the caring spectacles of celebrity, corporate philanthropy and NGO campaigning has authorized, set in motion and naturalized – if not outright deified – the continuing neoliberalization of the life and death of others in humanitarian aid and environmental programmes.

We need to ask ourselves if this neoliberalization of aid and its scaling up is really the only way forward, or if this forecloses the emergence and imagination of alternative socio-technical arrangements and infrastructures aimed at improving the human condition. Specifically, the market-driven rationale that informs the mission of these charities raises questions, and so do the ideological mechanisms that morally enable a rich Western entity to operate in the Global South – often represented as powerless and hopeless – thus replicating a highly problematic form of colonial thinking that is not conducive to much-needed structural change.[97] As I have outlined, there are analogies around being religious, being charitable and making sacrifices. Caring for others, and caring about the water crisis, demands a sacrifice. Global water charities constantly remind individuals – and consumers – that they are the ones that need to make a sacrifice to help solve the water crisis. After all, what is a donation, if not a sacrifice?

---

97 I. Kapoor, *Celebrity Humanitarianism: The Ideology of Global Charity*, London: Routledge, 2013.

3

Sacrifice: #WaterCrisis – Utopias and Relational Sacrifice

Aquaman to the rescue

The superstar actor Jason Momoa, of *Aquaman* fame, was handed the Nature Baton by the UN Special Envoy for the Ocean, Peter Thomson, who also designated Momoa the UN Environment Programme's (UNEP) Advocate for Life Below Water. Commenting on this appointment, UNEP's executive director, Inger Andersen, said that 'with a huge audience of engaged fans, we believe that Jason can move ocean considerations into the hearts and minds of citizens and business leaders to promote this urgency and action'.[1] This happened during the major UN Ocean Conference, held to mobilize 'global support for an ocean besieged by the triple planetary crisis of climate change, biodiversity loss and rampant pollution'.

The fact that celebrities are so deeply enmeshed in high-level humanitarian circles is both a cause and a symptom of the spectacularization of environmental action, global developmentalism and charity campaigns in late capitalism. Just as major private companies use celebrities as spokespeople to foster their marketing communications strategy – a phenomenon known as celebrity endorsement or celebrity branding – today it is common practice for international

---

1 UNEP, *UNEP Names Jason Momoa Official Advocate for Life Below Water*, press release, 27 June 2022.

organizations and governments to seal partnerships with celebrities and benefit from their fame. Indeed, in early 2022 I talked about the celebritization of aid with a senior UN humanitarian affairs officer, who was critical of it but also acknowledged its importance: 'Currently, the situation is dramatic. Funding is not growing (or maybe just a little), but the "needs" are growing exponentially. So, we needed to come up with something to leverage more resources.' But when did this happen?

The rise of celebritized forms of global humanitarianism and charity work can be traced back to the late 1980s with the formation of the Band Aid charity supergroup and the first Live Aid concerts. However, the phenomenon gained real momentum in the twenty-first century, led by the global charity work of entertainment stars and the related sale of celebrity-endorsed products, the rise of philanthrocapitalism and the work of NGOs such as Save Darfur and Médecins Sans Frontières. Celebrities, for their part, are eager to help, also because celebrity humanitarianism, Kapoor observes, is not always altruistic; rather, it is a self-serving promotion of the celebrity 'brand' that tends to advance consumerism and the reach of corporate capitalism. Today, charity is, according to the insightful observation of Jo Littler, part of the job description and an established hallmark of what it means to be a celebrity.[2] This is a function of the moral authority afforded to individuals as famous, wealthy and influential media celebrities, but also of the moral authority claimed by celebrities (and their media enablers, fans and charity organizations) through their work in charity and philanthropy.

Scholars such as Lilie Chouliaraki have critiqued the ways that celebrities and celebrity-fronted charity as a process have worked through this moral authority to create a 'theatricality

---

2 J. Littler, *Radical Consumption: Shopping for Change in Contemporary Culture*, London: McGraw Hill Education, London, 2008.

of solidarity' through a marketized 'post-humanitarianism'.[3] For Chouliaraki, the media performances of humanitarian celebrities create a 'humanitarian imaginary' through 'the personifying power of celebrity, the enchantment of the rock concert or the professional witnessing of the [celebrity] journalist, so as to confront us with the spectacle of distant sufferers as causes that demand our response'.[4] Empathy for distant others on the part of the audience is put to work through a 'moralistic education' by celebrity performances of authentic caring for others. Here, using Chouliaraki's example of Angelina Jolie, is an 'entrepreneurial confessionalism': the individualized emotional responses of celebrities (for example, tearing up upon witnessing poverty) are embedded in but also replicate the spirit of capitalism activated through the 'therapeutic value' of a 'gestural economy' of audience empathy and the contemporary political economies of charity.[5]

Further work has conceptualized humanitarian and caring celebrities through the multiple tropes that they articulate, such as the 'earth mother' in the form of Angelina Jolie, the 'ambassador' in Harrison Ford, the 'guru' in Jane Goodall and 'ordinary' celebrity activists such as Greta Thunberg.[6] One of the key tropes here is that of the 'white saviour', such as Diane Fossey, Bono, George Clooney, Madonna

---

3 L. Chouliaraki, 'The Theatricality of Humanitarianism: A Critique of Celebrity Advocacy', *Communication and Critical/Cultural Studies* 9: 1 (2012), pp. 1–21; *The Ironic Spectator: Solidarity in the Age of Post-humanitarianism*, New York: John Wiley and Sons, 2013.

4 Chouliaraki, *The Ironic Spectator*.

5 C. Rojek, '"Big Citizen" Celanthropy and its Discontents', *International Journal of Cultural Studies* 17: 2 (2014), pp. 127–41.

6 C. Abidin et al., 'The Tropes of Celebrity Environmentalism', *Annual Review of Environment and Resources* 45 (2020), pp. 387–410; L. A. Richey and D. Brockington, 'Celebrity Humanitarianism: Using Tropes of Engagement to Understand North/South Relations', *Perspectives on Politics* 18: 1 (2020), pp. 43–59.

and Nicholas Kristof, concerned with 'saving' Africans and African nature in the often less-than-hidden guise of mediagenic neo-colonialism.[7] In much of this work, scrutiny of the celebrification of humanitarianism and international development points to the proto-religious overtones of celebrities, their performances and the market-based, neoliberal mechanisms through which the political economies of charities are designed to function. In fronting for causes, celebrities become the 'high priest' of charities – both of their own and the causes they represent – given their ability to direct the gaze, intentions and actions of audiences through this theatricality of humanitarianism.[8] These high priests are moralizing pedagogues of the 'right' feelings (sympathy, empathy, care) and the 'right' actions (donation, conscious consumption, sacrifice) that frame problems and solutions to global poverty and environmental crises through the growing 'canon' of neoliberalized problem analysis and marketized solutions. They bear witness by doing tours to Africa and other 'needy' places to diagnose problems and share 'successes' of, for example, microloans for bore wells and improved toilet facilities that are then reported back to us through TV shows, YouTube videos, social media campaigns and news coverage.

Celebrity high priests also bear witness to their own 'conversion' to how bad the problem is, and what, then, is the best solution (for example, microfinance, Fair Trade, female entrepreneurialism). This is often done in the company of – and with the help of – ancillary 'lesser priests' in the advisors and development experts explaining problems and solutions to celebrities and their audiences. Humanitarian celebrities 'confess' their feelings, their own 'sins' and the poverty they have borne witness to in order to construct an authentic, emotional expertise that feeds their high priest persona. At

---

[7] K. Mathers, '"Mr Kristof, I Presume?": Saving Africa in the Footsteps of Nicholas Kristof', *Transition* 31: 107 (2012), pp. 14–31.
[8] Chouliaraki, *The Ironic Spectator*.

the same time, these confessions transfer a kind of power to audiences to become their own version of the high priest. This is all facilitated through the words of celebrities in the form of oft repeated urgings that 'your donation will support these communities in accessing clean water and sanitation' and the accompanying 'good deed' images of celebrities with the people and environments they have helped.

In addition to speaking to their 'congregation' – their audiences and fans – through media appearances, they also speak, as Bono has often put it, on behalf of this congregation to policymakers and other powerful elites.[9] Thus, celebrity high priests work to take both elites and fans with them into redemption through sacrifice, care and empathy, all monetized through donations, 'correct' purchases of 'helping' commodities and, sometimes, policy change or other elites' personal wealth. Indeed, through belief in these very often white saviours and the solutions they put forward, personalized redemption is near at hand for those willing to listen, contribute and believe.[10] Here, the public and often viral words and deeds of celebrity high priests facilitate our own redemption as paved through our own donations and the psycho-emotional, personalized 'saving' of ourselves in the 'saving' of those now with clean water and sanitation facilities. And, more often than not in this contemporary era, high priests are, as Nicole Aschoff would put it, the literal 'prophets' of fetishized neoliberal, technocratic solutions of philanthrocapitalism and consumer choice – and their accompanying political economies – they put before us as the solutions to global poverty and international development.[11]

---

9 D. Brockington, *Celebrity Advocacy and International Development*, London: Routledge, 2014.

10 M. T. Boykoff and M. K. Goodman, 'Conspicuous Redemption? Reflections on the Promises and Perils of the "Celebritization" of Climate Change', *Geoforum* 40: 3 (2009), pp. 395–406.

11 N. Aschoff, *The New Prophets of Capital*, London: Verso, 2015.

Much of the moral authority and power of celebrity high priests is spent on selling the public a neoliberal vision of charity and philanthropy that is very often concealed behind celebrity. Over time, celebrity-fronted philanthropic campaigns have begun to develop their own forms, logics, intentions and outcomes that more often than not eschew structural change.[12]

To return to Water.org and WaterAid, and their reliance on celebrity campaigns, there are two kinds of campaigns worth isolating: the first is structured around celebrity-marketed advocacy campaigns for conscious and ethical consumption, and the second focuses on celebrity-fronted fundraising campaigns. The former work to create and deploy revenue through shopping and consumer choice, spurred on by celebrity voices or modelling of 'good' shopping behaviour. The latter often have a different set of political economic goals through the deployment of celebrities' own economic capital, capital from corporate sponsorship and other, pre-existing, fundraising programmes, for example, donations to the Red Cross or funding streams from the UN. Oxfam and many NGOs include both political economic forms by allowing consumers to purchase products that support humanitarian and environmental causes, while at the same time they are supported by donations and other charity work. These 'upstream' caring, affective economies are then tied directly to the on-the-ground, 'downstream' political economies that caring celebrities, NGOs and foundations work to set up to support their humanitarian efforts through, for example, the building of schools in poor communities, supplying medicines, providing refugee support or, as in this case, supporting better access to clean water and sanitation.

---

12 N. Farrell, 'Introduction: "Getting Busy with the Fizzy" – Johansson, SodaStream, and Oxfam: Exploring the Political Economics of Celebrity Activism', in *The Political Economy of Celebrity Activism*, London: Routledge, 2019, pp. 1–18.

Middle- and upper-class shopping choices of the 'congregation' support the humanitarian and development aid of the far-flung poor, all at the behest of some of the world's richest and most powerful megastars. Lisa Ann Richey and Stefano Ponte[13] refer to this more generally as 'Brand Aid', whereby humanitarian, development and environmental aid is literally branded into consumer products such as those on the approved shopping list of Bono's Product (RED) campaign or the likes of Fair Trade goods, shoes and diapers.[14] This is truly the ideology of neoliberal capitalism laid bare in the harsh light of Giorgio Agamben's 'bare life': life and death of the poor and marginal are a function of the choices made by the global middle and upper classes for coffee, handbags and baby buggies, corporate largesse and (often rather cynical) CSR ploys, as well as the individualized choice of celebrity high priests to 'speak out' about particular environmental and humanitarian crises.[15]

That celebrities bring economic capital in the form of strategic partnerships with corporate entities, which promote market-based solutions and reinforce structural inequalities, is aided by the creation of their own organizations from which to speak to their congregations. Celebrity-fronted political economies of philanthropy and humanitarianism

---

13 L. A. Richey and S. Ponte, *Brand Aid: Shopping Well to Save the World*, Minneapolis: University of Minnesota Press, 2011; 'Brand Aid and Coffee Value Chain Development Interventions: Is Starbucks Working Aid Out of Business?', *World Development* 143 (2020), 105193.

14 M. Goodman, 'Reading Fair Trade: Political Ecological Imaginary and the Moral Economy of Fair Trade Foods', *Political Geography* 23: 7 (2004), pp. 891–915; A. Brooks, *Clothing Poverty: The Hidden World of Fast Fashion and Second-hand Clothes*, London: Zed Books, 2015; R. Hawkins, *A New Frontier in Development? The Use of Cause-related Marketing by International Development Organisations*, *Third World Quarterly* 33: 10 (2012), pp. 1783–801.

15 G. Agamben, *Homo Sacer: Sovereign Power and Bare Life*, Stanford: Stanford University Press, 1998 (or. ed. 1995).

facilitate the moral authority of a particular neoliberal political economy of sacrifice through the efforts of celebrity high priests and the donations of their congregations. Care for others and environments is materialized in not just the bodies, media appearances and cultural capital of celebrities, but also the associated products, donations, goods, elites, consultants and corporations that circulate in a political economy of affect and concern that then touches down in the places 'most in need' of support, aid, capital and/or market creation. Put another way, the political economies of sacrifice created by the caring spectacles of celebrity, corporate philanthropy and NGO campaigning have authorized, set in motion and naturalized – if not outright deified – the continuing neoliberalization of the life and death of others in humanitarian aid and environmental programmes.

Today celebrities do not only endorse and care about a specific social or environmental cause. They also tell us *what* to do and *how* to do it, and their voice has become increasingly authoritative, even though they are not necessarily experts on the matter at hand. And we – citizens, donors and consumers – have largely normalized this mechanism. These narratives and messages have gained legitimacy. We are no longer surprised when we see a celebrity taking on a cause. We might be intrigued by it, sometimes amused, and often we want to hear more, and perhaps join their cause.

Social water crisis?

Would the 'global water crisis' have the same social and political prominence that it has today without the presence of images of droughts, sun-baked fields of cracked earth, women carrying large jerrycans along a dirt path in an undefined countryside, rusting wrecks of ships in the Central Asian Aral Sea or of children drinking dirty brown water straight from a puddle? In his study of the visualities of humanitarianism, Fuyuki Kurasawa explained that images of a crisis can be interpreted as social actants. The simple act of viewing them constitutes persons into audiences that feel responsible to ease the suffering of those portrayed as victims: 'Seeing means knowing and, in turn, an obligation or compulsion to "do something" and help.'[16] And indeed, we are all an audience to the water crisis, even though not everyone is exposed to it in the same way.

Consider, for example, the exceptional drought that gripped southern Europe in the summer of 2022, which, according to the Copernicus Emergency Management Service – an information programme managed by the European Union – is possibly the worst ever experienced in the continent. Italy, and in particular the northern agricultural plain fed by the Po River, had to face its worst drought in seventy years due to an early heatwave and prolonged poor rainfall. In June, several shipwrecks from World War II resurfaced as the Po River reached low levels, and the Italian government declared a state of emergency. Farmers struggled to water their orchards, some municipalities started to rent portable desalination units, and Pietro Salini, the CEO of the Italian company Webuild – one of the largest construction companies in the world, particularly with regard to dams, hydroelectric plants and other hydraulic infrastructures – said that the best and quickest way out of this for Italy was to build several large desalination plants, all to be built within a development project called Acqua per la Vita (Water for Life).[17] As rumours about imminent

---

16 F. Kurasawa, 'How Does Humanitarian Visuality Work? A Conceptual Toolkit for a Sociology of Iconic Suffering', *Sociologica* 9: 1 (2015), p. 2.

17 Webuild Group, *Webuild CEO Pietro Salini: Invest in Desalination to Resolve Quickly, Structurally Italy's Drought Problem*, press release, 1 July 2022.

water rationing made their way into Italian political and popular debate, some commentators and journalists rightly put the blame on decades of underinvestment in Italy's water network, which has a loss rate of 42 per cent nationally. Unsurprisingly, however, individual responsibility and the role of individual action gained a prominent place within the debate, and the media started to share advice on how everyone could and should make a difference from home. The private sector keenly jumped in as well: for instance, the bottled water giant S.Pellegrino – owned by the Swiss multinational company Nestlé – put forward a decalogue detailing how and why it is important not to waste water.[18] Rather obviously, the decalogue did not mention the numerous conflicts between Nestlé and local communities across the world over the draining of natural water supplies for bottling, but we will return to this aspect in the next chapter. Again, and similarly to what I outlined in the introduction, citizens are being addressed primarily as consumers, and not as part of a political community.

While the 2022 drought in Italy and southern Europe is certainly serious, it is also an extraordinary event, at least in its severity. Droughts are indeed a recurrent feature of the European climate, and even though their frequency is increasing due to anthropogenic climate change, they tend to occur once every five or six years.[19] The drought-induced water crisis in Europe is therefore a geographically situated and context-specific event, and this is the reason why, in spite of the alarmism and dramatic tone used to describe it, when the first rain comes most people will forget about it, and act surprised when it happens again in a few years. Then there is the structural water crisis, one that affects, and that

---

18 S.Pellegrino, *Come e perché è importante non sprecare acqua*, San Pellegrino website.

19 European Environment Agency, *Meteorological and Hydrological Droughts in Europe*, 23 March 2020.

has been affecting, several countries in the Global South, particularly in Africa and in Asia, and to which we are yet to find a solution. There are humanitarian consequences to this water crisis, both because it has been happening for decades and because the countries experiencing it are more vulnerable to it and are less prepared to face it compared with, for instance, southern European countries.

We – citizens of the developed world – seem to have normalized the fact that some undefined countries in a generic 'developing world' are just 'not doing well', 'lagging behind' or 'need help', as if this is just the way things are. Arguably, this comes with an implicit acknowledgement that the developing world is facing a series of structural problems, rather than context-specific ones. And yet, the hegemonic narrative around the 'global water crisis' foregrounded by global water charities and the corporate world does not really pay heed to the reasons behind this structural problem. Rather, it focuses on localized solutions that tend to provide an immediate 'fix'. Such a line of thinking also underpins a large part of the fundraising campaigns led by global water charities, which tend to propose localized solutions, rather than advancing structural radical change.

As noted earlier, a donation to a charity can be interpreted as a fetish, a means through which we establish a relationship with a distant place – or an inaccessible celebrity – without doing so fully. The fetishization of aid is accompanied by a simplification, or even a disavowal, of the multiple and complex relations that have established the need for that same donation, for example for bore-well technology that allows access to local, clean water. On one hand, aid turns into a post-political arrangement; on the other hand, the fetishized donation becomes immaterial and happens in a utopia, where utopia is understood in its literal sense as a non-place. And what is a donation, if not a sacrifice? While nowadays a donation is interpreted by some as a way to get tax relief, the meaning of a donation is profound, and it is rooted in the deeply spiritual concept of sacrifice. Here, the work of the French philosopher Georges Bataille and his notion of *dépense* (translated in English into 'expenditure' but also 'excess') is useful to explain that a sacrifice is meaningful only when painful, only when it leads to a loss, to the act of giving up something that matters.[20] Drawing mostly on the case of the *potlatch* – a gift-giving feast practised by Indigenous peoples of North America – Bataille explains that a cult

> demands a bloody waste of sacrificial men and animals. A sacrifice is no more, in its etymological meaning, than the production of sacred things. Sacred things are constituted by a loss: in particular, the success of Christianism must be explained through the value of the defamatory crucifixion of the son of God, who leads human anxiety towards a limitless representation of loss and degradation.[21]

To ensure its own survival, the capitalistic religion manipulates the notion of sacrifice, normalizing a system in which we are asked to make small-scale sacrifices – such as donating tiny amounts of money or buying a beer chalice to help a water charity, as we will discuss later – and diminishing the real meaning of sacrifice. Capitalism produces and demands a series of domesticated and symbolic sacrifices that neutralize the possibility for real sacrifices to be offered or even discussed.[22] In both capitalism and religion, certain activities and words are prohibited and become taboos. One of the main capitalist taboos is arguably that of the gift, the notion of gratuity.

---

20 G. Bataille, *La part maudite: précédé de la notion de dépense*, vol. 20, Paris: Éditions de Minuit, 1967.

21 Ibid., p. 29.

22 L. Bruni, 'The Capitalistic Religion: Old Questions, New Insights', in L. Bouckaert, K. J. Ims and P. Rona (eds), *Art, Spirituality and Economics*, London: Springer, 2018, pp. 159–69.

As Luigino Bruni illustrated,[23] elaborating on René Girard's book *La violence et le sacré*, traditions of ancient totemic civilizations, whereby the untouchable object of the taboo – often an animal – was sacrificed annually, consumed and eaten to take possession of its force and keep social violence at bay, are reproduced by the logic of capitalism.[24] In this logic, the idea of the gift is evoked, teased, used and consumed, once a year or a few times per year, so that it is deprived of its force. This happens through volunteering initiatives led by employees (often in partnerships with charities), fundraisers to help the poor, occasional donations and so on. Such forms of symbolic philanthropy and subdued gifts give companies a way to sacrifice a tiny fraction of their revenues to reproduce the appearance of a real gift and keep real gratuity at bay. The culture of the homeopathic gift, the principle that a tiny fraction of poison is used to cure and neutralize larger amounts of that same poison, is typical of the mechanisms through which tiny gifts can make us immune and neutralize a real gift. It is only when we are conscious of the fact that these gifts are merely symbolic that we can start questioning them and start thinking about real material and structural change.

Collective responsibility, individual action

The parasitic relationship between capitalism and religion feeds undeniably on the manipulation of the notion of sacrifice. The concept of sacrifice, as discussed, is a profound one, and is radically attached, following Bataille's discussion of the *potlatch*, to the notions of loss and pain. And while the

---

23 L. Bruni, 'Sul confine e oltre / 9. Violiamo il grande tabù', *Avvenire*, 18 March 2017.
24 R. Girard, *La violence et le sacré* [*Violence and the Sacred*], Paris: Éditions Bernard Grasset, 1972.

example of the *potlatch* is a rather extreme one, it is clear that a sacrifice, in its literal sense, has to make the experience of giving uncomfortable.

Take the case of Abraham, for instance, who was ready to sacrifice his son Isaac as a burnt offering to God.[25] Or consider the rich young man, told by Jesus that he must give all his fortune to the poor if he were to enter the kingdom of God, thus hinting at the interrelations between privilege, sacrifice and justice.[26] Capitalism and liberal philanthropy have, to an extent, embraced this approach, at least on paper. In 2020, Darren Walker – the president of the Ford Foundation – referred to the activist Luis Miranda Jr. to note that:

> 'If it doesn't hurt … then you're not giving enough.' This conception of sacrifice is profound: Only when it is uncomfortable, even painful to give of ourselves, only when your life makes a meaningful shift away from something you otherwise might want, do you know that you are giving beyond your own benefit.[27]

These words deserve some praise, even though, as Karen Ferguson – a scholar who has thoroughly studied the Ford Foundation – observed, the Ford Foundation is concerned with the disruption of the levers of inequality, but it does not really try to eliminate 'either the privilege or the levers'.[28] More generally, there is a certain dissonance between theory and praxis in the workings of liberal philanthropic organizations, and the demanded sacrifice tends to be hardly painful or uncomfortable.

Consider, for example, WaterAid's campaign 'Give it up for Taps and Toilets', a fundraising toolkit aimed at raising

---

25 Genesis 22, Bible.
26 Mark 10: 17–31, Bible.
27 D. Walker, 'The Hard Work of Hope', *Ford Foundation*, 22 February 2020.
28 K. Ferguson, 'The Perils of Liberal Philanthropy', *Jacobin*, 26 November 2018.

awareness about the water crisis through the involvement of average citizens in the UK and the US.[29] In this campaign, volunteers are asked to turn their 'self-sacrifice into a fundraising frenzy!':

> What can't you live without? Decide what you're going to give up and how long you're going to do it for. Whether you're giving up beer for a year, or giving up chocolate for a month, remember that it needs to be enough of a challenge for people to sponsor you. The joy of this idea is it takes minimal organisation, just bucket loads of willpower![30]

And if giving up chocolate for a month is too much of a sacrifice, there is another option: 'donate your big day'.

> Don't fancy giving up tea or Keeping up with the Kardashians, but have a big celebration coming up? Why not give up your birthday, wedding or another notable life event, and ask for donations to WaterAid instead of gifts. If you're having a party to celebrate, we can send you some WaterAid resources to jazz up the place. Get in touch at events@wateraid.org and we can arrange to send you WaterAid balloons, bunting and leaflets.[31]

Everybody can thus promote their sacrifice, make an event out of it, and raise money for taps and toilets. Social networks and sharing with friends are an important element of these initiatives, and volunteers are encouraged to keep their supporters updated on their progress via social media or the Just Giving online platform. People share their achievements and campaigns, explaining the reasons behind their sacrifice, which, for most, consist of a runathon, a long-distance walk or run to raise money for WaterAid (or any other charitable cause), but sometimes are also a way to remember a loved

---

29 WaterAid, *Give it up for Taps and Toilets: Your Fundraising Toolkit*.
30 Ibid.
31 Ibid.

one. Those participating show a genuine will to help, and in their personal profile they explain that they have been exposed – mostly through social media – to how dire the water crisis is, and they thus decided to 'do something' about it, with runathons tending to be that 'something'. There is not a general rule for how a runathon works, but most of the time it follows this scheme: someone with a passion for sports sets a goal for themselves, such as running from London to Edinburgh or swimming around Hyde Park's Serpentine, all while raising money for a charitable cause. Each day, the person will share updates on progress towards their target and explain why they are raising awareness about a cause. Runathons are both a public event and an individual project that revolve around sharing with others a personal sacrifice, or in some cases, even a pseudo-martyrdom, as in the case of runathon superstar Mina Guli.[32]

Guli is an Australian businesswoman who in 2012 founded the Thirst Foundation, a non-profit dedicated to 'increasing awareness, driving stakeholder urgency and delivering meaningful action on the resource that connects everyone – water'.[33] Like other high priests, or priestesses, of charity, Guli has an inspiring story:

> A swimming pool prank gone wrong would prove to be an accident that would change the course of my life. I hurt my back so badly that doctors told me I'd never be able to run again. Given my lifelong antipathy towards sport, this might have been good news. I could have used this diagnosis as an excuse to sit on the sofa and eat pizza. Instead, I saw it as an opportunity – to redefine my limits.[34]

---

32 E. Stutchbury and S. Kirkham, 'Mina Guli Is Running 200 Marathons in a Single Year to Open Our Eyes to an "Invisible" Crisis', ABC News, 27 April 2022.
33 See thirstfoundation.org.
34 Mina Guli personal website, minaguli.com.

Guli thus started swimming, which then led to biking, and eventually to running. In parallel, her career as a lawyer picked up. She began working on privatizations and infrastructure investment, and eventually joined the WEF's community of Young Global Leaders, where she was introduced to what she calls 'the problem of "invisible water"', that is, the water embedded in food and other commodities, based on the concept coined by the late Tony Allan known, in the scholarly community, as 'virtual water'.[35] Shocked by the seriousness of the water crisis, Guli started to take on a number of ultra-running challenges to raise awareness of the issue. In 2016 she completed the Seven Desert Run, a series of forty marathons over forty-nine days and across seven continents. Thanks to this achievement, she made it into Fortune's list of the fifty greatest leaders in the world, alongside Pope Francis, Bono, Jeff Bezos, John Oliver and Christine Lagarde. As Fortune's website explains: 'On March 22, World Water Day, she completed her 1,048-mile journey. "Never seen a better example yet of #gobigorgohome," tweeted a fan in Hong Kong.'[36] In 2017 it was the Six River Run (forty marathons in forty days), which left her unable to walk and forced her to spend months rebuilding the muscles in her legs.

In 2019 it would have been 100 marathons in 100 days, accompanied by the #RunningDry and #EveryDropCounts hashtags. But her body had other plans. By day sixty-two, a stress fracture on her femur forced her to drop the challenge. She initially insisted on carrying on, but then abandoned to avoid permanent damage to her leg. A carefully crafted video posted on her Twitter page shows a badly limping Guli in tears, eventually being admitted to hospital in a wheelchair. But not everything was lost. Guli's supporters completed

---

35 T. Allan, *Virtual Water: Tackling the Threat to Our Planet's Most Precious Resource*, London: Bloomsbury, 2011.

36 'The World's 50 Greatest Leaders', *Fortune*, 2016.

the 62nd marathon on her behalf, prompting the official account of the Global Water Partnership South America to leave the following comment: 'You deserved it! At last you generated a much greater awareness movement. Sorry that the cost is really high', thus emphasizing Guli's self-sacrifice for the water cause.

In 2020, speaking at the UN-Water High Level Political Forum, Guli remarked:

> Over the last 4 years, I have put my body on the line to not only bring awareness to our global water crisis, but to drive action to solve it ... When I set out last year to run 100 marathons in 100 days for water it was the biggest, boldest way I thought I could show what 100% commitment on this issue looked like. But I fell short. I broke after 62 marathons. I thought I had failed ... But then my community rallied around me. People around the world stepped up to run in my place, to share their own water stories and make their own commitments. And I realised: individually we can make an impact, but together we change the world.[37]

Guli unveiled her latest project in 2022: running 200 marathons across 200 countries – with the finish line set in New York for the UN Water Conference in March 2023 – to raise awareness about the 'massive global water crisis', 'move 200 per cent faster', get 200 companies to join the Thirst Foundation Six for 6 framework and get support from 200 million people across the countries she visits. A thought leader with high-level political support and influence, Mina Guli is also well connected to most major global water charities and businesses operating in the water sector. With her tendency for martyrdom and her tokenistic calls for community involvement, Guli embodies the modern neoliberal

---

37 Mina Guli CEO of Thirst remarks UN-Water-HLPF to launch the SDG-6 Global Acceleration-Framework.

water activist and Good Samaritan, ultimately driven by a largely self-centred mission and ethos, and a carefully crafted communications campaign which seems more concerned with presenting Guli as an illuminated visionary rather than with the explanation of the complexities and contradictions of the contemporary water crisis. But while Guli became a celebrity through her commitment to solve the water crisis, usually it is the other way around, as well-established celebrities get directly involved in the 'global' quest.

Matt Damon's toilet strike – a donation campaign developed by Water.org – provides a fitting example. In 2013, Damon decided to launch a toilet strike to raise awareness about the water crisis. As he explained in an interview:

> Yes, I went on a toilet strike last year. We're trying to come up with, uh, it's very hard to message with this stuff, particularly around water, you know, it's like a marketing person's nightmare because it's such a complex issue. It's hard to boil it down into kind of a little digestible sound bite. And so we're starting to toy around with the idea of using comedy to try to, you know, create some viral videos.[38]

Celebrities endorsed the toilet strike, and the likes of Richard Branson, Olivia Wilde and Bono pledged 'not to go to the bathroom until everyone in the whole world has access to clean water and sanitation'.[39] Supporters were asked to use the hashtag #strikewithme to share funny news about the strike and their sacrifice, such as 'I feel there is a troll living inside me' or 'I haven't peed or pooped in four days and I feel great'.[40] Damon urged people to 'join the millions of famous people who've already joined my strike, and

---

38 CNBC, 'Matt Damon's Toilet Strike', YouTube, 2 June 2014.
39 Water.org, *Celebrities Endorse the Toilet Strike*, press release, 22 May 2013.
40 Water.org, 'Jason Bateman, Jessica Biel and Josh Gad Support the Strike!', YouTube, 13 March 2013.

remember, if you don't use the toilet, you're a celebrity!'[41] The whole campaign is intentionally comical and revolves around scatological humour, which essentially leads to a spectacularization of the act of 'going to the bathroom'. Yet what is also interesting here is the fact that donors can become celebrities by proxy if they follow their lead, thus assuming that people desire to become celebrities, and that a donation thus becomes a means to this end. The required sacrifice (not using a toilet) is entirely symbolic: people are not really required to act on their pledge, but rather, they are asked to act as brand advocates for Water.org, and, additionally, urged to make a monetary donation. Such a meaningless sacrifice is only a pale imitation of the struggle faced daily by the 4 billion people (half the world's population) who do not have access to safely managed sanitation in their homes.[42] The toilet strike plays into the absurdity of the real that encapsulates the late capitalist, Anthropocene response to ecological, humanitarian and health crises. The matter of fact (that half the world's population does not have access to sanitation) is fully decontextualized and reduced to a marketing operation and a self-promotion exercise.

Buy yourself out of the crisis

The high priests of charity produce economic capital through the donations of their congregations, but also by means of strategic partnerships with corporate entities that promote market-based solutions and conscious consumerism, a phenomenon that Lisa Ann Richey and her research team aptly defined as the commodification of compassion. Overall, the two largest global water charities, Water.org and WaterAid, rely on donations from individuals quite differently.

---

41 Ibid.
42 UNICEF, WHO, *1 in 3 People Globally Do Not Have Access to Safe Drinking Water*, 18 June 2019.

Table 1: Overview of WaterAid UK's main sources of income in the period 2017–2021. Source: WaterAid UK's annual reports and financial statements.

Year	Total Income (million GBP)	Individuals	Corporate and trust supporters	Other WaterAid member countries	Governments and other institutions
2020–1	89,5	50.3	12.0	8.9	17.8
2019–20	91,3	48.7	17.1	11.9	12.9
2018–19	91,4	45.9	21.1	12.8	11.1
2017–18	83,4	44.7	16.9	11.9	5.8

Table 2: Overview of Water.org's main sources of income in the period 2018–2021. Source: Water.org's annual reports and financial statements.

Year	Total Income (million US$)	Individual supporters	Foundations, corporations and other organizations	Investment and other
2021	37.5	3.4	32.0	0.6
2020	19.3	6.4	12.1	0.8
2019	24.6	7.6	16.0	0.9
2018	30.6	3.9	25.8	0.8

As shown in Tables 1 and 2, while donations from individuals tend to make up more than half of WaterAid's total income, Water.org mostly relies on donations from (and partnerships with) private foundations and corporations. This confirms the different ethos of the two charities that was discussed in the previous chapter. On one hand, WaterAid tends to portray itself as a collective charity, one that originated through a coordinated effort of the British water companies, and that favours, at least on paper, citizen-driven participatory approaches. On the other hand, Water.org clearly favours an individualistic approach to solving the water crisis, as evidenced by the strong role played by its founders, White and Damon, and its predilection for private investors and equity funds. Nevertheless, partnerships with large multinational corporations are crucial for both charities. WaterAid partners with, among others, H&M and the H&M Foundation, the Heineken Foundation, the banks HSBC and AXA XL, and Belu Water (which we will cover in the next chapter). Likewise, Water.org has conducted several large fundraising campaigns over the years, with, among others, the IKEA Foundation, Danone Aqua (which we will also discuss in Chapter 4), the PepsiCo Foundation, Inditex and Stella Artois.

As mentioned in this book's opening, the case of Stella Artois – a beer brand owned by the world's largest brewer, Anheuser-Busch InBev SA/NV – is particularly revelatory of how capitalism manipulates the notions of gift and sacrifice, rendering them empty and using them to expand its reach. In March 2019, Water.org and Stella Artois celebrated World Water Day with the #PourItForward campaign. As Stella Artois's press release explains:

> Stella Artois and Water.org co-founders Matt Damon and Gary White are rallying America to 'Pour It Forward®' by choosing Stella Artois to help end the global water crisis – and they're challenging some famous friends to spread the word, starting with Sarah Jessica Parker ... It's easy to get involved. For a limited time, every Stella Artois helps provide access to clean water for someone living without it:
> - Every Chalice gives access to 5 years of clean water for one person in the developing world.
> - Every 6-pack gives access to 6 months of clean water for one person in the developing world.
> - Every 12-pack gives access to 12 months of clean water for one person in the developing world.
> - Every pour (or bottle) sold at bars and restaurants gives access to 1 month of clean water for one person in the developing world.[43]

---

43 Stella Artois, 'Sarah Jessica Parker Joins Stella Artois and Water. org to "Pour it Forward®" and Help End the Global Water Crisis', PR Newswire, 22 January 2019.

Consumers were thus asked to spread the word using the #PourItForward tag on social media, and Stella Artois pledged to trigger product donations of US$3.13 for every limited-edition Stella Artois chalice sold in the United States. Stella Artois also purchased a thirty-second spot during the 2018 Super Bowl (Stella Artois, 2018) for an estimated price tag of US$5 million (Gharib, 2018), representing a fifth of Water.org total expenditures in that year, in which Damon encouraged viewers to buy a limited-edition chalice to help end the global water crisis. Rather than being aimed at making a real difference and helping humanity solve the water crisis, such an emotional and spectacular call to arms contributes to raising the profile of both Stella Artois and Water.org and, by implication, Matt Damon performing in his celebrity high priest role, while also boosting celebrity-led ethical consumption among its growing congregation.

Similarly, Water.org and the Levi's® brand celebrated World Water Day 2012 with the 'Go Water<Less' campaign, to encourage people to adopt a 'Water<Less™' lifestyle.[44] The campaign – which understandably does not make any reference to the fact that the garment industry is one of the most polluting in the world, and that it takes 3,781 litres of water to make a pair of jeans – invited people to visit the Levi's® campaign site to complete several challenges that raised awareness on how to reduce individual water consumption.[45] Each challenge completed unlocked a number of Water.org 'WaterCredits' and participants were introduced to the water-conscious Water<Less™ jeans collection, so that 'globally conscious consumers can learn more about the global demand for clean water and the

---

44 Water.org, *Levi's Water<LESS Campaign*, press release, 22 March 2012.

45 For more information refer to the UN Environment Programme website.

small actions they can take each day to use less and give more'.[46]

Going back to toilets, their dichotomous role as both a basic need and a business opportunity is also detectable in the 'More than a Toilet' campaign launched in 2018 by Water.org together with Harpic, a company producing a toilet bowl cleaner that also happens to be the leading toilet cleaner in India – a crucial country for Water.org – and Lysol, a US-based brand of cleaning and disinfecting products who also have a large share of the Indian market. Harpic and Lysol claim to have created a social experiment in which people are denied access to the toilet. The companies do not provide any details of the social experiment, which appears to be a marketing operation revolving around a one-minute video that, at the time of writing, has over 8 million views on YouTube.[47] Alice Moore, Harpic's Global Category Director, welcomed the initiative, adding:

> We are very grateful to be working towards Water.org's efforts to end the water and sanitation crisis around the world. It is exceptionally shocking to see just how many people around the world are still living without access to basic sanitation and the worrying effect this has on their health, safety and education. We are so passionate about our 'More than a toilet' campaign and the positive contribution we at Harpic can make through raising awareness and fundraising, alongside RB's US$1 million donation.[48]

While of course a welcome effort and impressive if taken out of context, Reckitt's donation of US$1 million is a meaningless fraction of Reckitt's (Harpic's and Lysol's parent company) annual revenues for that year, which amounted to more than US$14 billion. Hardly a sacrifice.

---

46 Water.org, *Levi's Water<LESS Campaign*.
47 Harpic UK, 'Toilet Access Denied, How Would You React? Harpic & Water.Org Find Out …', YouTube.
48 Ibid.

Beyond these fundraising campaigns, cause-related marketing – and not necessarily celebrity-fronted – has become increasingly pivotal in the quest to solve the water crisis. Water.org leads the way here as well. Their 'shop to support' platform features a growing list of partnerships with several dozen companies ranging from coffee producers to water filtration bottles, including, somewhat ironically, an electronic toilet spray that releases odour-blocking essential oils when someone sits on the bowl. The act of buying any of these products also triggers a small donation to Water.org, whose impact is summarized by means of unverifiable claims such as: if you buy this 'you will provide one person with access to safe water for more than two years', which work towards the fetishization of both commodities and donations.[49] There is no relationship among people in this transaction: I buy a toilet cleaner, and as a result, someone far away in an undefined poor country might get a tap or a toilet, so I might as well buy several toilet cleaners – and encourage my friends to do the same – as they are helping to solve the water crisis.

The act of giving and the notion of sacrifice – and the political economies built around this sacrifice in donations and purchasing decisions – are thus crucial for these campaigns. But they have also come a long way from their forebears, as discussed by René Girard, turning into their pale imitation.[50] Under this logic, a donation is no longer a *gift* – one of the main taboos in capitalism alongside the notion of gratuity – but rather another opportunity to buy goods and engage in consumption patterns, all the while normalizing and sustaining the capitalist imperative of growth that helps partner companies expand their businesses. Individuals who have made a donation are prompted to share this on social media to encourage others to do the same,

---

49 See water.org.
50 Girard, *La violence et le sacré*.

making the experience of giving both public and implicitly reciprocal.

The seemingly win-win scenarios portrayed by charity-led marketing are not that different from those used by large equity funds to justify investments on the 'global water crisis'. In late 2020 it became possible for the first time to trade water on Wall Street through the futures market, which has traditionally hosted commodities such as oil, oranges or gold.[51] Trading in the futures market literally means that buyers commit to purchasing a commodity at a predetermined future date and price, regardless of market conditions at the set date. In this case, CME Group launched the Nasdaq Veles California Water Index futures (NQH2O) to help investors 'manage the price risk associated with the scarcity of water in the largest water market in the US', California. NQH2O sets a price in US$ to buy a fixed amount of water corresponding to 'the volume of water required to cover one acre of land (43,560 square feet) to a depth of one foot, equivalent to 325,851 gallons'. As the Nasdaq overview of the NQH2O illustrates, this is a clearly speculative financial market based on water availability. During periods of drought the NQH2O index rises and with it the market price for water. In periods of wet hydrological conditions, the price for water decreases. Putting a price to water, according to Nasdaq, is inevitable:

> It would be nearly impossible to overstate the vital role that water plays in our lives. It stands at the top as an unnegotiable requirement for life. It is critical to agricultural production, manufacturing energy production and even transportation. Despite this, advancements in other commodities markets have far outpaced those related to water, leaving market participants as well as consumers, exposed to considerable water price risk

---

[51] A. Tobin, 'Could Trading Water on the Stock Market Actually Be Good for the Environment?', Euronews, 17 May 2021.

... With NQH2O, Nasdaq offers its expertise in benchmark production to the betterment of markets supporting our most precious natural resource.[52]

The NQH2O is all about crisis management and price certainty. It does not provide a way to better manage water resources, or make sure that water does not get polluted or lost on the way because of leaky pipes and decades of underinvestment. On the contrary, it can be argued that NQH2O bets on the assumption that water is bound to become an increasingly scarce resource, and access to it will be less predictable. With it comes an implicit acceptance – almost a welcome – of the failures of water governance and of the inevitability that the water crisis is going to get worse. And indeed, we can expect to see other water future markets opening in other areas of the United States, and generally in other parts of the world where water is scarce and in high demand.

So, it is now largely agreed upon that we have a water crisis, that we should do something to help (or protect ourselves from it), and that in order to do this, we need to invest in innovative solutions. Take, for instance, the Fidelity® Water Sustainability Fund (FLOWX), an investment fund that capitalizes in companies that are 'helping drive greater efficiency, extending the lifecycle, improving infrastructure, and developing disruptive technologies' in the water sector.[53] The fund is based on the assumption that the world is facing critical water shortages, and that the situation is going to get worse due to a booming global population, decaying infrastructure and climate change. The fund's webpage explains that because of these issues, 'we're seeing increasing demands from governments around the world asking for help – in particular, help from private companies' that are seemingly

---

52 Website of the Nasdaq Veles Water Index.
53 Fidelity, *Global Water Crisis: Investing in Water*, 12 June 2023.

those best placed to help since 'public budgets and spending are under pressure'. Such a combination – helpless governments and a worsening water crisis – make for a good investment opportunity, since revenues in the water sector could grow 4 to 6 per cent each year, thus confirming the results of a study commissioned by Water.org in 2016.[54] And just like Water.org looks at solving the water crisis as a means to expand towards new large and untapped markets, FLOWX sees promise in countries such as India or China where the water crisis is likely to trigger new capital expenditures in the decades to come. As Janet Glazer, Fidelity's lead analyst writes: 'My optimism is grounded in a very realistic view on the global crisis that is in front of us today, making the call to action more important now than it ever has been.'[55]

Capitalism can thus look at the future with confidence. Humans, however, should be more wary.

Conclusions

Is running a marathon – or giving up chocolate for a month – going to solve the water crisis? Are these meaningless sacrifices really going to make a difference, or rather, are they sustaining the illusion that we are making a difference? The fact that investment firms speculate on the water crisis comes as no surprise. Crisis is, after all, one of the structural foundations of capitalism. But the above discussion raises important questions about the contradictory relationship between capitalism and charity, which is strongly shaping the very nature of giving in the water sector. Global water

---

54 L. Pories, 'Income-enabling, not Consumptive: Association of Household Socioeconomic Conditions with Safe Water and Sanitation', *Aquatic Procedia* 6 (2016), pp. 74–86.

55 Fidelity, *Global Water Crisis*.

charities reproduce, and are produced by, the ideas and processes that have more generally determined the neoliberalization of the water crisis. Rather than being a social or natural resource, water has fully become a commodity, and as such, it naturally flows into commercial and (micro) financial circuits. If, on one hand, private equity funds lure investors with the promise of a good economic return on a monetary investment, on the other hand capitalist charity does the same (as in the case of the water equity fund), while also stretching and manipulating the notion of sacrifice to gratify donors/investors as they engage in yet more consumption patterns. While water charities, and their high priests, incessantly remind us that we do have a water crisis, they fail to explain why we have one, disregarding the complex structural dynamics and unequal power relations that have caused it. Capitalism has appropriated the water crisis, and feeds on it, and this is evident through the alliances and relationships triggered by this dynamic.

4

Corporate Social Redemption: Common Water, Bottled Profit

Whenever and wherever someone reaches for a plastic bottle, there is a very good chance that that bottle contains water. Bottled water consumption has grown steadily in recent decades, and while our sick planet is in the midst of multiple water crises, there certainly is not one for the bottled water sector. It might even be argued, as recent research has shown, that as the combined effects of urbanization and climate change are exacerbating water scarcity, bottled water is increasingly emerging as *the* quick fix. Urbanization seems unstoppable. In 2014 the UN noted that, for the first time, more than half of the world's population was living in urban areas,[1] and, as the late Mike Davis observed, cities will account for all future world population growth.[2] In parallel, bottled water consumption also seems unstoppable. In 2016, bottled water officially surpassed carbonated soft drinks to become the largest beverage category by volume in the United States.[3] This is impressive, even more so if we consider that per capita consumption of bottled water in the United States (170 litres per year) pales in comparison with that of other countries like Thailand (216 litres) or

---

[1] UN, *More Than Half of World's Population Now Living in Urban Areas, UN Survey Finds*, 2014.
[2] M. Davis, 'Planet of Slums', *New Left Review* 26 (2004).
[3] 'Americans Drank More Bottled Water Than Soda in 2016', Reuters, 10 March 2017.

Mexico (282 litres).[4] The situation is similar in Europe, where packaged water accounted for half of the total sales of non-alcoholic drinks in 2019, with countries like Italy, Germany and Portugal leading the way.[5] The trend is even stronger in the Middle East and Africa, where bottled water makes up over two thirds of the soft drinks market, and demand is predominantly driven by lack of access to clean and safe water, and rapid urbanization.[6]

But urbanization, together with decades of underinvestment in water infrastructure and climate change, only tells part of the story of bottled water, whose fortunes across continents are at the essence of the neoliberal venture, and of cultural capitalism. While in poor countries or in places hit by an exceptional water crisis (as in the case of the city of Flint, in Michigan) bottled water is often the only source of clean and safe water; in rich countries – where potable tap water is available at very low or no cost – the surge in its consumption is largely a cultural phenomenon.[7] As Gay Hawkins, Emily Potter and Kane Race explain in their book *Plastic Water*, much of the success of bottled water is due not only to the fact that plastic has become inexpensive and very easy to produce, but also because our contemporary water ontologies are deeply entrenched in the discourses of hydration, wellness, living 'on the go', and more generally in the appreciation of the therapeutic effects of mineral water.[8] To exemplify this, we can draw upon two watery vignettes.

---

4 Statista, *Per capita consumption of bottled water worldwide 2020, by leading countries*, July 2021.

5 Natural Mineral Waters Europe, *Statistics*.

6 Euromonitor International, *Bottled Water in the Middle East and Africa*, September 2020.

7 R. Wilk, 'Bottled Water: The Pure Commodity in the Age of Branding', *Journal of Consumer Culture* 6: 3 (2006), pp. 303–25.

8 G. Hawkins, E. Potter and K. Race, *Plastic Water: The Social and Material Life of Bottled Water*, Cambridge: MIT Press, 2015.

Consider, for example, the case of the French company Evian – one of the most well-known mineral waters in the world – and its branding strategy. The company was founded in 1859 in the Alpine city of Évian and was originally called Société anonyme des eaux minérales de Cachat, after Gabriel Cachat, the owner of the garden where the water flowed. At that time, Europe was in the midst of its golden age of thermal towns, as thermotherapy (the idea that bathing in hot water and breathing its vapour had therapeutic effects) was becoming increasingly popular among intellectuals, aristocrats and the bourgeoisie. It is within this context that in 1865 the city of Évian changed its name to Évian-les-Bains (in French, *les bains* refers to a spa town) to attract more visitors and develop its thermal business. Four years later, in 1869, the company also changed its name to Société anonyme des eaux minérales d'Évian-les-Bains; this is significant, as thereafter, it became common practice to name a mineral water after the toponym of its source, and not of its owner. Initially, the Evian water was marketed for its medicinal properties as a thermal water and was therefore primarily sold in pharmacies. In the meantime, in the early 1900s, the Alps – that until then had been largely perceived as a forgotten and dangerous place, or at best as a natural barrier – started to emerge as the epitome of all that was 'natural', regenerating, pure and pristine. Large European industrial cities certainly offered money and opportunities to their citizens, but these came with increasingly polluted air, filthy streets and a high population density. Thus, Europe's wealthiest added the Alps to their list of top destinations to reconnect with nature and breathe pure air. With the Alps being the new sanctuary of wellness, Evian started to be marketed as the purest and lightest of waters, ideal for everyday consumption but also to prepare a formula feed.[9]

---

9 'Saga Evian', *Revue des marques* 33 (2002).

As consumers were having different concerns, Evian tried to assuage them. And so, in the 1960s, drinking Evian became a means to regenerate oneself and fight the wear and tear of modern life, something that today we would colloquially call detox. In the 1980s, alongside rising concerns about polluted water and mistrust of municipal hydraulic infrastructure, Evian portrayed itself as a cleaner alternative to tap water. Then, in the late 1990s, as life expectancy in France grew from sixty to eighty years old in just three decades, Evian became the water that helped seniors to 'postpone their retirement until they turned ninety'. During those years, and following the company's acquisition by the French multinational food giant Danone, Evian became a global brand, and its sales went from 8 million litres in 1901, to the nearly 2 billion litres of 2022, most of it outside of the French market. The invention of the polyethylene terephthalate (PET) bottle in the 1970s, which replaced the more fragile and rigid polyvinyl chloride (PVC), was a game changer. Evian launched several new water bottle formats, and in the 2000s the company established numerous partnerships with celebrities and high-fashion designers ('a unique opportunity for a match made in hydration heaven'), further establishing itself within the market of luxury brands. In 2018, Evian's partnerships with the Italian influencer Chiara Ferragni to release a limited-edition bottle of water stirred controversy in Italy, as politicians and consumers found its €8 price tag to be immoral and unethical. And of course, the Ferragni bottle quickly sold out. Evian thus went full circle from being a therapeutic water sold in pharmacies to being a container of the complex – and often contradictory – assemblages and constellations of the late capitalist era.

If we move to the other side of the Alps, in Northern Italy, another spa town – San Pellegrino Terme – gave its name to another iconic bottled water, S.Pellegrino. The history of the company Sanpellegrino S.p.A is similar, in many ways, to that of Evian. S.Pellegrino was first sold in the late nineteenth century, and the water was at the time renowned for its therapeutic effect. The company experienced a steady growth during the last century, and in 1998 it was bought out by the Swiss multinational company Nestlé (Danone Bottled and Nestlé Waters are respectively the first and the third largest bottled water companies in the world). Today most of S.Pellegrino's sales are made outside of Italy, and in 2022 the S.Pellegrino Group sold 3.5 billion bottles.[10] Significantly, and unlike Evian, S.Pellegrino made a name for itself as a flag-bearer for its country of origin, and therefore for 'authentic Italian values'. According to the company, drinking S.Pellegrino is equal to 'living in Italian'. Its advertising campaigns tend to portray a utopian idea of the Italian way of living, which is largely idealized and relies heavily on Federico Fellini's imaginary of the *Dolce Vita*. Take, for example, S.Pellegrino's 2022 TV commercial. In it, the American actor of Italian descent Stanley Tucci visits San Pellegrino Terme and notes:

> Ever have a moment you wish you could perfectly capture and bottle forever? ... Actually, how convenient, perfect moments like this can happen all the time. With the right, people food, and drink, like a moment here in the beautiful mountains, just outside of San Pellegrino Terme where water travels through the rocks on a 30-year journey, that enriches it with minerals ... Moments spent in the glamorous casino of San Pellegrino Terme. Or, moments of community friendship and passion, Italian virtues ... These virtues are alive in the streets of this town and in the water that shares its name.[11]

---

10 B. Pekic, 'Italy's Sanpellegrino Reports Turnover Of €892m in 2020', *European Supermarket Magazine*, 26 May 2021.

11 S.Pellegrino, *Virtù italiane, Stanley Tucci protagonista della nuova pubblicità di S.Pellegrino*, press release, 10 February 2022.

Beyond being a status symbol, S.Pellegrino stretches the territory of Italy beyond its material borders, providing those who drink it with a localized *sense of place* that is reassuringly immutable and at the same time fully globalized and cosmopolitan. This is quite a leap, considering that, after all, we are still talking about water.

So, bottled water is also a cultural phenomenon. But how is this connected to the 'global' water crisis? And is this an addition to the problem, or a potential solution?

The state of bottled water

As bottled water enjoys its golden era, its consumption is now fully normalized and has become a habit. A central element in the growth of bottled water is the commodification[12] of a resource that until recently was seen and treated as a public good, and which was seen and treated, as Karen Bakker outlined, as one of the 'material emblems of citizenship'.[13] The privatization of water resources has followed an irresistible trend. As Michael Goldman observed in 2007, 'as recently as 1990, fewer than 51 million people received their water from private water companies, and most water customers were in Europe and the United States', but the number rose to 460 million people in the early 2000s, with high-growth areas in Africa, Asia and Latin America.[14] Bottled water is following a similar path, and in this context it is appropriate to talk about the *capture* and *enclosure*

---

12 R. Pacheco-Vega, '(Re)theorizing the Politics of Bottled Water: Water Insecurity in the Context of Weak Regulatory Regimes', *Water* 11: 4 (2019), p. 658.

13 K. Bakker, *Privatizing Water: Governance Failure and the World's Urban Water Crisis*, Ithaca: Cornell University Press, 2010.

14 M. Goldman, 'How "Water for All!" Policy Became Hegemonic: The Power of the World Bank and its Transnational Policy Networks', *Geoforum* 38: 5 (2007), p. 790.

of what was once a public resource. Bottled water thrives, among other things, on mistrust of municipal or national water supply, and this usually plays out into a simple antinomy: bottled water is pristine and *natural*, while public water is untested and potentially polluted. This can be traced back to broader concerns about the public sector and its ability to deliver a service of a similar quality to that offered by private companies, which tend to be perceived as more efficient and regulated. Loss of trust in the public sector is a structural problem, and it is the result of decades of underinvestment in hydraulic infrastructure and privatization policies across the globe, further exacerbated by widely reported disasters like the Flint water crisis or the contamination of the Yamuna River in India. As Raul Pacheco-Vega has argued:

> Governmental failures to provide safe drinking water through local water utilities, poor networked infrastructure for water delivery throughout urban centers, rural and peri-urban areas, powerful marketing campaigns, regulatory failures and capture of local governments on the part of multinational corporations, a taste for healthy hydration through highly portable liquids, and a shift in norms where consuming bottled water has become somewhat of a cultural norm despite its negative environmental effects are all factors that have contributed to the emergence and sustained growth of the global bottled water industries.[15]

Bottled water is convenient, portable and relatively inexpensive if compared with other soft beverages. But if we look at it as *water*, and not as a soft drink, bottled water turns out to be quite expensive, at least in absolute terms. If, for instance, a litre of bottled water in a European supermarket costs on average €0.30, tap water in Europe costs, on average, €1.50 per cubic meter (equivalent to 1,000 litres), and it is thus around 200 times cheaper than bottled water. And while

---

15 Pacheco-Vega, '(Re)theorizing the Politics of Bottled Water'.

paying one or two euros for a bottle of water might not be particularly taxing for the average European citizen, things are quite different elsewhere in the world, also because bottled water follows the rules set by the market. This was evident, for example, in the Mexican city of Monterrey, which in 2022 was gripped – like many other parts of the world – by a severe drought. While taps ran dry for citizens (the infamous Day Zero became reality), the fact that Coca-Cola (who also sells bottled water), Heineken and other firms still had access to water caused public outcry, and led the Mexican government to ask private companies to show solidarity and give their water to citizens.[16] As bottled water was in high demand, its price tripled between May and July, costing nearly as much as gasoline. Against this backdrop, those who could not afford to buy bottled water were forced to drink unclean brackish water distributed by *pipa* trucks operated by the municipality, while state security forces had to patrol a nearby dam to prevent the theft of water.[17]

'Glocal' water

While this may sound like a dystopian scenario – and to many extents it is indeed one – conflicts between bottled water companies and local communities are quite common. As already discussed, Niagara Bottling – the largest supplier of private label bottled water in North America and an important partner of Water.org – has been involved in numerous controversies, conflicts and legal battles with local communities about water pumping in aquifers across the United States. At a general level, two main problems

---

16 Comición Nacional del Agua, Gobierno de México exhorta a industriales *y agricultores de Monterrey a ceder agua temporalmente para abasto a la población*, 28 June 2022

17 L. Perlmutter, '"It's Plunder": Mexico Desperate for Water while Drinks Companies Use Billions of Litres', *Guardian*, 28 July 2022.

can arise when it comes to granting water pumping rights to a private company. The first occurs when a company can extract water in areas where water is scarce or already over-exploited, as in the above-mentioned case of Mexico. The second is when water becomes scarce as a result of the water withdrawals carried out by a bottled water company. But scarcity is a contentious term when we talk about water, and for that matter, when it comes to the environment more generally.[18] Lyla Mehta has, for instance, illustrated how water scarcity can be both absolute and manufactured, remarking on the 'need to analyse water scarcity at two levels: one, at the discursive level where scarcity is "constructed" and two, at the material level as a biophysical problem where it is lived and experienced differently by different people'.[19] Anything, including water, can become suddenly scarce depending on what you want to do with it, and there are many cases in history in which a perceived water scarcity led to the construction of large infrastructural projects such as mega dams or giant canals.[20] In the case of bottled water, there is a clear clash between the local and the global scale, a symptomatic condition of the processes behind 'glocalization'.[21]

18 Refer for instance to G. Kallis, *Limits: Why Malthus Was Wrong and why Environmentalists Should Care*, Stanford: Stanford University Press, 2019.

19 L. Mehta, 'Whose Scarcity? Whose Property? The Case of Water in Western India', *Land Use Policy* 24: 4 (2007), p. 661.

20 As I wrote elsewhere (Menga, *Power and Water in Central Asia*), the Soviet Union carried out its hydraulic mission in Central Asia in the 1940s and 1950s, with the plan of making 'mad rivers sane' and bringing water to the desert, where water was indeed rather scarce and not sufficient to grow a water-intensive crop such as cotton. To reverse these 'natural' constraints, the Soviet Union built more than sixty canals to divert water from the Amu Darya and the Syr Darya rivers, including the Kara-kum Canal, one of the longest (1,400 km) irrigation canals in the world, that tapped into Amu Darya River to bring water in the Kara-kum desert in Turkmenistan.

21 Glocalization, as Erik Swyngedouw noted ('Globalisation or "Glocalisation"? Networks, Territories and Rescaling', *Cambridge*

This is evident in the case of the tiny Osceola Township.[22] Nestlé Waters began its operations there in 2008, when it constructed a water pipeline in Osceola County to transport water from a local well to a factory located in Evart, Michigan, where its Ice Mountain brand of water was bottled. Nestlé's existing permit allowed the company to pay, overall, US$200 a year to the state of Michigan to pump around 500 million litres of water.[23] In 2016, due to increasing global demand for its bottled water, Nestlé asked permission to build a booster station to pump 400 gallons of water (equivalent to 1,514 litres) a minute, 150 more than the 250 a minute it was already pumping. Osceola Township residents, who had already seen a decline in the water flow of the Twin Creek River, protested and rejected the plan, saying that it did not comply with its zoning laws. Nestlé's Natural Resources Manager for North America remarked that – based on studies commissioned by Nestlé itself – there had been no measurable changes to the streams, and wherever those changes occurred, these were not due to Nestlé, but rather to some dams on those streams.[24] Nestlé subsequently sued Osceola Township. A court ruled in its favour, saying that water was essential for life and that bottling water was an 'essential public service' that met a broader demand. However, in 2019 an appellate court reversed the decision, explaining that:

Review of International Affairs 17 (2004), p. 1), 'refers to the twin process whereby, firstly, institutional/regulatory arrangements shift from the national scale both upwards to supra-national or global scales and downwards to the scale of the individual body or to local, urban or regional configurations and, secondly, economic activities and inter-firm networks are becoming simultaneously more localised/regionalised and transnational'.

22 A township in Michigan with a population of 1,000 people, mostly retirees.

23 J. Glenzain, 'Nestlé pays $200 a year to bottle water near Flint – where water is undrinkable', *Guardian*, 29 September 2017.

24 AFP News Agency, 'Tiny Michigan Town in Water Fight with Nestlé', YouTube, 4 February 2018.

We agree with the trial court's observation that water is essential to human life, as well as to agriculture, industry, recreation, science, nature, and essentially everything that humans need. However, the trial court went on to conclude that because selling bottled water at a profit supplies a public demand somewhere, it constitutes a 'public service'. A 'public service' means 'the business of supplying a commodity (as electricity or gas) or service (as transportation) to any or all members of a community' or 'a service rendered in the public interest'. *Merriam-Webster's Collegiate Dictionary (11thed)*. The first definition would not be unreasonable if the sale of bottled water approximated a public utility subject to regulation by the Public Service Commission or a similar entity. The second definition would not be unreasonable if plaintiff was primarily in the business of supplying bottled water to areas that lacked any other source of potable water. Plaintiff's commercial operation satisfies neither understanding of a 'public service'. Furthermore, other than in areas with no other source of water, bottled water is not essential. The trial court erred in effectively concluding that because water is essential, the provision of water in any form, manner, or context is necessarily an 'essential public service'.[25]

This is noteworthy, not only because the court set an important precedent regarding the award of water pumping rights, but also because it clearly called out the business of bottling water for being, unsurprisingly, a business. Of course, a priori this is not an issue, but it can become one when said business is portrayed as an essential public service, since it is neither essential, nor public.

To contextualize this, in 2006, when asked about his views on water as a commons, the CEO of Nestlé, Peter Brabeck-Letmathe, observed that:

---

25 *Nestlé Waters N. Am., Inc. v. Twp. of Osceola*, No. 341881 (Mich. Ct. App. Dec. 3, 2019).

Water is, of course, the most important raw material we have today in the world. It's a question of whether we should privatize the normal water supply for the population. And there are two different opinions on the matter. The one opinion, which I think is extreme, is represented by the NGOs, who bang on about declaring water a public right. That means that as a human being you should have a right to water. That's an extreme solution. The other view says that water is a foodstuff like any other, and like any other foodstuff it should have a market value. Personally, I believe it's better to give foodstuff a value so that we're all aware it has its price, and then that one should take specific measures for the part of the population that has no access to this water, and there are many different possibilities there. I am still of the opinion that the biggest social responsibility of any CEO is to maintain and ensure the success and profitable future of his enterprise. For only if we can ensure our continued long-term existence will we be in the position to participate in the solution of the problems that exist in the world.[26]

While Brabeck-Letmathe is not explicitly denying that water should be a human right – and this is how his words have generally been interpreted by his critics – it is nevertheless clear that for him, and implicitly for his company, water is a raw material, and thus a commodity and a product, but certainly not a public service.[27] At this point it is worth taking a step back to revisit what happened during the 2nd WoWF, which took place in 2000 in The Hague, the Netherlands. As usual, the forum brought together representatives of the private sector (the majority) including private water providers, over 140 governments, some NGOs, and development practitioners. The forum culminated in the adoption of the

---

26 Transcript from Zereau Drinks, *Nestlé CEO Peter Brabeck over Drinkwater*.

27 P. Muir, 'The Human Rights and Wrongs of Nestlé and Water for All', *The National*, 28 November 2013.

'World Water Vision: Making Water Everybody's Business', a document that is largely seen as a key step in the global water privatization agenda.[28] As Wendy Barnaby – a science journalist who wrote an event report for the academic journal *Medicine, Conflict and Survival* – observed, 'one of the main recommendations of the Vision is that fresh water must be recognized as a *scarce commodity* [emphasis added] and managed accordingly. In other words, water consumers must pay the full cost of the water services they use.'[29] Based on the 'Vision', investment for water services needs to come increasingly from the private sector, whereas governments should focus mostly on environmental protection and subsidies to the poor. In short: profit and profit-making activities are the prerogative of private companies, and corporate social responsibility is a public social responsibility. Nestlé, together with other water companies, played an important role in the drafting of the 'Vision'. A forum report published in a 2000 issue of the *Corporate Europe Observer* provides a snapshot of the *ethos* of the event:

> In the showroom area of the conference (the World Water Fair), corporations such as Nestlé, Suez Lyonnaise des Eaux, Unilever, and Heineken showcased their efforts to promote sustainability and water efficiency, while their CEOs addressed the assembly demanding that water be recognised as an economic good rather than as a human right. In fact, all the Forum rhetoric focused on human 'needs' as opposed to a concept of human 'rights', consistent with the World Water Vision in which the concept of water as a human right does not appear. This emphasis away from discussion of human rights complimented with the

---

28 Later published as a book: W. J. Cosgrove and F. R. Rijsberman, *World Water Vision: Making Water Everybody's Business*, London: Routledge, 2014.

29 W. Barnaby, '2nd World Water Forum', *Medicine, Conflict and Survival* 16: 3 (2000), p. 326.

following Ministerial Meeting declaration's recognition of water as an economic good proved to be a major ideological victory for the corporations seeking inroads into the water market.[30]

The above is interesting because it underlines how the WoWF has consistently been, over the years, a key driver of global water deregulation and privatization.

Let us now return to Osceola Township. On one hand, there is a tiny local community that feels that its water is getting increasingly scarce. On the other hand, we have a transnational company that says that the water they are allowed to pump is not enough to meet global demand. Indeed, water is arguably scarce in Osceola Township, but only as a result of Nestlé's operations in the area. If Nestlé were to be taken by some sort of revolutionary spirit and decide not to follow a business model inspired by the imperatives of growth and the generation of profit and, say, settle at a certain volume of sales, the water in Osceola Township would not by any means be scarce, and Nestlé would not feel the need to ask permission to build a booster station. However, since selling bottled water is highly profitable, Nestlé has a legitimate interest in increasing its production. The increase of water withdrawals and the search for new water pumping sites is part of capitalism's structural tendency towards environmental degradation, the expansion of Jason Moore's commodity frontier, which refers to those previously uncommodified spaces into which capital must expand and commodify nature to perpetuate itself.[31] Through

---

30 'And Not a Drop to Drink! World Water Forum Promotes Privatisation and Deregulation of World's Water', *Corporate Europe Observer* 7 (2000).

31 J. W. Moore, 'Sugar and the Expansion of the Early Modern World Economy: Commodity Frontiers, Ecological Transformation, and Industrialisation', *Review* 23: 3 (2000), pp. 409–33; 'The Capitalocene Part II: Accumulation by Appropriation and the Centrality of Unpaid Work/Energy', *Journal of Peasant Studies* 45: 2 (2018), pp. 237–79.

this inherently contradictory process, the fully global and globalized bottled water market clashes with the interests of a small local community. And while Nestlé points out that there is an objective need to pump more water in Osceola Township to provide a basic essential service to people living in an undefined elsewhere, the provision of this service leads to the denial of that same service where the company operates, and it creates a local water crisis to respond to an allegedly global one.

And if one might be inclined to think that this only happens in the United States – a developed country where water is relatively abundant – similar (and to many extents more contentious) cases occur elsewhere, including in countries particularly vulnerable to water crises. In Pakistan, for example, Nestlé's Pure Life has repeatedly been accused of depriving local communities – particularly poor and disadvantaged groups – of potable water.[32] Likewise, in Manderegi, Nigeria, where in 2016 Nestlé opened a large Pure Life bottling plant, residents have been struggling with chronic water shortages and contaminated water supplies, and Nestlé's corporate social responsibilities projects did not deliver the expected results.[33] Furthermore, while bottled water can often be presented as a cleaner alternative to tap water, the case of Nigeria – where research has revealed that sachet and bottled water is largely contaminated by heavy metals and microbes such as Escherichia coli, Enterococcus faecalis and Pseudomonas aeruginosa – shows that this is not necessarily the case.[34]

---

32 N. Rosemann, 'Drinking Water Crisis in Pakistan and the Issue of Bottled Water: The Case of Nestlé's "Pure Life"', *Actionaid Pakistan* 4 (2005), p. 37; R. Regenass, 'Poisoning The Well? Nestlé Accused of Exploiting Water Supplies for Bottled Brands', *Business and Human Rights Resource Centre*, January 2012.

33 A. Abba, 'How Nestlé Nigeria Contaminates Water Supply of Its Host Community in Abuja', *ICIR*, 28 April 2019.

34 O. J. Ajala et al., 'Contamination Issues in Sachet and Bottled

Drowning in plastic

The invention of the PET bottle in the 1970s has been a game changer for bottled water, and more generally, for humanity. Whereas we now take it for granted, PET is most certainly a material of wonder. It provides us with a cheap and yet very effective way of storing and protecting our foods and liquids so that they can be easily portioned, transported and consumed. Plastic has become crucial for food packaging. For example, Ian Cook et al. have brilliantly told the story and assemblages behind the global papaya trade, and the importance of the green plastic crates used to store them.[35] Single-use plastic packaging is also a transient material, one that seamlessly enters and exits our daily routine. We might buy a soft drink, keep it in our bag for a few hours, drink it and then the plastic bottle that unassumingly stored it for us becomes waste. Some of this plastic waste – only 9 per cent globally, according to the OECD – then ends up in a recycling centre, 22 per cent is mismanaged, 49 per cent goes into a landfill, and the rest is incinerated.[36] Plastic consumption has quadrupled over the past thirty years, and greenhouse gas emissions from plastic production – which account for roughly 5 per cent of global emissions – are rising, since plastic is increasingly produced in coal-based economies.[37]

What is most concerning, at present, is plastic pollution from single-use plastic packaging; it is one of the most pressing environmental issues of our time. As the OECD points

---

Water in Nigeria: A Mini-review', *Sustainability Water Resources Management* 6: 112 (2020).

35 I. Cook, *Follow the Thing: Papaya, Antipode* 36: 4 (2004), pp. 642–64.

36 OECD, *Plastic Pollution Is Growing Relentlessly as Waste Management and Recycling Fall Short, Says OECD*, press release, 22 February 2020.

37 L. Cabernard et al., 'Growing Environmental Footprint of Plastics Driven by Coal Combustion', *Nature Sustainability* 5: 2 (2022), pp. 139–48.

out, most plastics in use today are virgin plastics, made from crude oil or gas, while recycled plastics only account for 6 per cent of total plastics production.[38] In 2019, out of the 460 million tonnes (Mt) of plastics produced globally, 6.1 Mt leaked into aquatic environments and 1.7 Mt flowed into oceans. Overall, there is an estimated 30 Mt of plastic waste in seas and oceans, and 109 Mt in rivers. This means that the leaking of plastic waste into the ocean will continue for decades, particularly because plastic waste is largely mismanaged, as evidenced by recent research showing that private water companies in the UK are the main source of microplastics – plastic particles smaller than 5 mm – pollution in UK rivers.[39] The Great Pacific Garbage Patch (GPGP) – a gyre of buoyant ocean plastic three times the size of France – is rapidly accumulating more plastic, reminding humans, as if that were ever necessary, that the Anthropocene is a *tangible* geological epoch.[40] Calling it a garbage patch, however, is counter-intuitive, as the GPGP does not really look like an island of plastic, but is rather a cloudy mix of microplastics and larger plastic waste, with much more debris (around 70 per cent of the total) lying in the ocean's depths.

But microplastics are not only floating in our rivers and oceans. Recent research has provided evidence that microplastics can now be commonly found in human blood, meconium, breastmilk and even human placenta, or 'plasticenta'.[41] While this is often used as a rhetorical figure,

---

38 OECD, *Global Plastics Outlook: Economic Drivers, Environmental Impacts and Policy Options*, Paris: OECD Publishing, 2022.

39 J. Woodward et al., 'Acute riverine microplastic contamination due to avoidable releases of untreated wastewater', *Nature Sustainability* 4: 9 (2021), pp. 793–802.

40 L. Lebreton et al., *Evidence that the Great Pacific Garbage Patch Is Rapidly Accumulating Plastic*, Scientific Reports 8 (2018), 4666.

41 A. Ragusa et al., 'Deeply in Plasticenta: Presence of Microplastics in the Intracellular Compartment of Human Placentas',

we can safely maintain that *Homo sapiens* has truly turned into a somehow downgraded form of a cyborg made of both plastic and organic matter. It is thus unsurprising that plastics and microplastics pollution is under close popular, but also political, scrutiny. Even the former Conservative British prime minister Boris Johnson took a stance against plastic, saying that recycling plastic materials 'doesn't work' and 'is not the answer' to threats to global oceans and marine wildlife, and instead 'we've all got to cut down our use of plastic'.[42] Johnson's strong position against plastic sparked criticism from environmental activists and recycling associations, who both remarked that we do indeed need to reduce our plastic use, but we also have to reuse and recycle it. The point is that plastic is too deeply entrenched in the consumption (and protection) of food, water and energy, and humanity is simply not ready to give it up anytime soon. And yet, recycling plastic is also expensive, largely inefficient and not economical: there are now thousands of different types of plastic and none of them can be melted down together. Furthermore, the more plastic is reused, the more toxic it becomes, and it soon can no longer be recycled into food-grade packaging.[43] Indeed, Boris Johnson does have a point. We have cheap and easy-to-produce virgin plastic, but we also have expensive and sometimes toxic recycled plastic, which inevitably ends up being only a fraction of the total plastic used in any given product. Both are bad for the environment. Plastic, like global warming, presents humanity with a seemingly unsolvable puzzle. We cannot live without

---

International Journal of Environmental Research and Public Health 19: 18 (2022), 11593; A. Ragusa et al., 'Plasticenta: First Evidence of Microplastics in Human Placenta', *Environment International* 146 (2021), 106274.

42 G. Bowden, 'Recycling Plastic Does not Work', BBC News, 25 October 2021.

43 Greenpeace, *Circular Claims Fall Flat Again*, 2022.

plastic, but plastic is also a threat to our lives as they are currently unfolding. A solution to such a problem can thus only result in numerous compromises and lesser-evil solutions. This does not mean that we should not be seeking a solution, even though this will be far from perfect. But we need to know that the invention of plastic has set in place an irreversible process, and there is no going back to life without plastic.

The bottled water business is clearly well aware of this, and a transition to plastic-free bottled water is out of the question, at least for the moment. As we will discuss later, most of the sustainability discourse propagated by bottled water companies revolves around the notions of recycling, water stewardship and sustainable water management, while the environmental consequences of plastic are largely ignored. In the unfortunate and rather grotesque case of Water.org's partner Niagara, a PET bottle is even brought forward as a more efficient alternative to nature-based 'packaging', since, for instance, a PET bottle comprises 2 per cent package and 98 per cent product, whereas an egg allegedly comprises 13 per cent package and 87 per cent product. But while eggshells do not contain microplastic, plastic bottles do. Recent research has shown that out of a total of 259 individual bottles from across 11 different brands and 27 different lots, 93 per cent showed some sign of microplastics contamination, including in brands such as Evian, S.Pellegrino and Nestlé Pure Life.[44] Interestingly, while microplastics contamination occurred in a brand of water that was packaged in both plastic and glass (Gerolsteiner), there was considerably less contamination within the water bottled in glass as compared to that packaged in plastic,

---

44 S. A. Mason, V. G. Welch and J. Neratko, 'Synthetic Polymer Contamination in Bottled Water', *Frontiers in Chemistry* 6: 407 (2018); Nestlé Pure Life had the highest contamination levels among all tested waters.

which indicates that water can be polluted at the source (and this might also happen to tap water), but a larger contamination occurs as a consequence of the packaging itself. And while the impacts of microplastics intake on human health are still unknown, we can safely say that if we want to reduce microplastics contamination – in both the environment and in our bodies – we should stop consuming bottled water, particularly when clean tap water is available.[45]

Bottled water is also fully global, and as such, it is extremely mobile. For example, 80 per cent of plastic water bottles consumed in the United States end up in landfills, but these landfills are not necessarily located in the United States. A study conducted in 2019 revealed that the United States sends 1 Mt of plastic overseas every year, and most of this goes to some of the world's poorest countries such as Bangladesh, Laos and Senegal, where waste is generally mismanaged and is either burned or thrown into the sea.[46] And beyond plastic as waste, plastic as a container of water is largely exported. These exports are not led by local needs or resource availability, but rather by cultural and social drivers. A dated and yet still relevant study conducted by UNEP in 2006 highlights the complexity of bottled water's patterns of global trade. France, Italy and China lead exports, while the United States, Germany and Japan emerge as the main importers, even though potable tap water is easily available. Moreover, major exporters – France, Germany and Belgium – are also major importers, and this further underlines the fact that bottled water is often not purchased

---

45 J. Nizamali, S. M. Mintenig and A. A. Koelmans, 'Assessing Microplastic Characteristics in Bottled Drinking Water and Air Deposition Samples Using Laser Direct Infrared Imaging', *Journal of Hazardous Materials* 441 (2023).

46 E. McCormick et al., 'Where Does Your Plastic Go? Global Investigation Reveals America's Dirty Secret', *Guardian*, 17 June 2019.

out of necessity, but rather as a non-essential product.[47] And clearly, transporting water long distances generates significant emissions of greenhouse gases (for example, CO_2 emissions from trade-related international freight amount to 30 per cent of all transport-related CO_2 emissions from fuel combustion), on top of the emissions triggered by the production of PET.[48]

We have come a long way from those Evian bottles sold in pharmacies across south-eastern France. Bottled water consumption is a fully normalized habit, but it is also a very controversial one which feeds on and nurtures some of the main dynamics at the core of the global water crisis: resource capture and accumulation; unequal power relations and the clash between the local and the global; the economic valuation and the commodification of nature; and the sidelining of local needs and wants.

There are thus many sins that bottled water needs to redeem. So how is this redemption sought?

Corporate Social Redemption?

The origins of Corporate Social Responsibility (CSR) are usually traced back to the work of Howard Bowen, an economist who published the book *Social Responsibilities of the Businessman*, which advocated for better business ethics and outlined the social responsibilities of the private sector.[49] CSR as a concept took flight in the 1970s, particularly in the Anglophone world, and it is now ubiquitous in

---

47 UNEP-GRIS Arendal, 'Total Bottled Water Consumption', in *Vital Waste Graphics 2*, 2006.
48 International Transport Forum, *The Footprint of Global Trade*, 2016.
49 H. R. Bowen, *Social Responsibilities of the Businessman*, Iowa City: University of Iowa Press, 1953.

the vision statement of almost every company worldwide. While its definitions are heterogeneous – as is common with highly debated and pervasive ideas – CSR generally refers to a company's responsibility towards both the communities and the environment in which it operates. This results in a series of economic, environmental, legal, ethical and even philanthropic responsibilities (but not obligations) that influence how a company tries to redistribute some of its own wealth.[50] If, on one hand, a private company might see in it a business opportunity to be developed through a business strategy, an NGO or any civil society organization might take it as an opportunity to hold a company accountable for its damaging environmental and social impacts. And when these pledges remain unfulfilled, CSR becomes *greenwashing*, 'the act of misleading consumers regarding the environmental practices of a company or the environmental benefits of a product or service' (see for instance the case of the above-mentioned 'egg-plastic bottle' analogy).[51]

Unsurprisingly, greenwashing is a widespread phenomenon. This is both because the environment has emerged as a key arena of political and social encounter and struggle, and also because as the socioecological crisis gets worse, the need to act becomes more compelling and the claims made by companies become more ambitious and wide-ranging. Consider, for example, the 27th UN Climate Change Conference of the Parties (commonly known as COP27). The NGO Global Witness conducted an analysis of all the delegates in attendance, finding that overall there were more representatives of the fossil fuel industry (636) than total delegates from the ten countries most impacted by climate change (Puerto Rico, Myanmar, Haiti, Philippines,

---

50 B. Carroll, 'The Pyramid of Corporate Social Responsibility: Toward the Moral Management of Organizational Stakeholders', *Business Horizons* 34: 4 (1991), pp. 39–48.

51 TerraChoice, *Greenwashing Report 2009*.

Mozambique, The Bahamas, Bangladesh, Pakistan, Thailand and Nepal).[52] Greenwashing has become such a widespread phenomenon that at COP26 in Glasgow, the UN Secretary-General António Guterres announced the establishment of a High-Level Expert Group to address net-zero commitments by non-state actors (particularly businesses, financial institutions, cities and regions). In 2022, the Expert Group published its first report, and the chair – Catherine McKenna, the former Minister of Environment and Climate Change of Canada – opened it by writing that 'it's time to draw a red line around greenwashing'.[53] The report thoroughly criticized greenwashing and provided a roadmap to bring integrity to the environmental commitments set by governments and the private sector, noting that companies cannot claim to be 'net zero' while continuing to build or invest in new fossil fuel supply or any kind of environmentally destructive activities. As McKenna observed, reiterating the sense of impending doom discussed in previous chapters, 'right now, the planet cannot afford delays, excuses, or more greenwashing ... too many of the net-zero pledges are ... little more than empty slogans and hype'.[54] So, if the UN are on it, greenwashing must be a big problem.

Research on greenwashing is consequently abundant, and scholars have extensively looked at how this phenomenon occurs in various sectors, studying its effects on consumer behaviour and the ways in which greenwashing unfolds on the ground. But while this research is, in its own way, compelling, it also somehow misses the point. Obviously – and

---

52 Global Witness, *Over 100 More Fossil Fuel Lobbyists Than Last Year, Flooding Crucial cop Climate Talks*, 2022.

53 UN High-Level Expert Group on the Net Zero Emissions Commitments of Non-State Entities, *Integrity Matters: Net Zero Commitments by Businesses, Financial Institutions, Cities and Regions*, 2022.

54 UN, *'Zero Tolerance for Greenwashing', Guterres Says as New Report Cracks Down on Empty Net-zero Pledges*, 2022.

similarly to what we have observed in the case of charity – CSR can be twisted and manipulated and turned into yet another money-making venture. As consumers are getting more concerned about the environment, businesses diligently show that they also *care*. I could easily challenge those reading this book to find a company that proudly lists all the ways in which their activities have led to environmental degradation, social marginalization and legal bullying. None will be found. Greenwashing is one of the structural symptoms of late capitalism, and as such, its occurrence is entirely normal (its absence, conversely, would be abnormal). What is more concerning is the fact that it seems to be easier to trust the sustainability agenda and recommendations set by transnational companies, CEOs, elite Good Samaritans, celebrities and marketing experts rather than listening to what climate scientists, geologists, hydrologists, conservationists, biologists and more have to say on the matter. It is clearly the former that set the tone and establish a mainstream discourse, not the latter.

It is for these reasons that instead of talking about Corporate Social Responsibility, it would be more accurate to refer to it as Corporate Social *Redemption*, with redemption understood as the process of freeing oneself from guilt or a debt. This allows us to give greater emphasis to the innate exploitative nature of the production-for-profit system, and of its need to constantly legitimise its actions. Since 'being sustainable' is arguably the most pressing social and cultural resolution of our times, the study of greenwashing can tell us a great deal about how capitalism constantly adapts and transforms itself in order to preserve its own existence. If we can understand how capitalism achieves this self-preservation, we can more effectively prevent its expansion. This compels us to further our understanding of what sorts of relations are forged (or avoided) through greenwashing in the corporate sector, and how charity – and more in general philanthrocapitalism – has become a crucial element in these profit-making activities.

When it comes to critically discussing the CSR initiatives launched by bottled water companies, however, research has so far been limited, besides the usual studies examining a correlation between CSR and consumer behaviours. Among the few studies that critically discuss the paradox of sustainable bottled water, Roberta Hawkins's and Jody Emel's article in *cultural geographies* examined ethically branded bottled water products, taking as a case study Ethos Water, a company with the mission of 'helping children get clean water'.[55] To do so, Ethos Water foregrounds the purchase of bottled water as a way of solving the 'global water crisis'. In what is a typical case of cause-related marketing, Ethos Water pledged to make a US$0.05 donation to a development organization or a charity – including WaterAid – for each bottle purchased (for approximately US$1.80). Echoing the point made earlier about *who* sets the agenda, the founders of Ethos Water explicitly stated their intention 'to educate Americans about the world water crisis since general awareness about the issue was "quite low"', even though this education ends up proposing simplified and inaccurate versions of the water crisis, with solutions that are 'wound up with marketing campaigns to sell more product and discourses that further legitimize the brand as an actor within global water governance practices'.[56] So, according to Ethos Water (now owned by Starbucks), bottled water is not only sustainable, but it also brings us one step closer to solving the water crisis.

But self-declared *ethical* bottled water companies are not the only ones who frame their product as sustainable.

---

55 R. Hawkins and J. Emel, 'Paradoxes of Ethically Branded Bottled Water: Constituting the Solution to the World Water Crisis', *Cultural Geographies* 21: 4 (2014), pp. 727–43.

56 Ibid., p. 739

As Catherine Jones, Warwick Murray and John Overton have illustrated, the luxury brand Fiji Water also claims to be sustainable, and it has indeed gained an international reputation as a supposedly clean and green product, thus collecting celebrity endorsements from the likes of Nicole Kidman and Barack Obama.[57] Fiji Water is sourced from an artesian aquifer on Fiji's main island, and is promoted as 'untouched' and 'unspoiled by the compromised air of the 21st century', something that strongly resonates with what we have seen earlier with Evian's marketing strategy. Yet the sustainability of Fiji Water has been questioned, and environmentalists think that its annual water extraction rate is excessive. The company also uses high-grade plastics – shipped from China – to bottle its water, relying on a plant that uses diesel-generated energy. The water is then shipped all over the world, and the final product is thus the result of intensive fossil fuel use. To offset its carbon emissions (and I might add, seek redemption), Fiji Water launched numerous reforestation projects, engaged in philanthropic initiatives, and started using recycled plastics, which resulted in accusations of greenwashing by environmentalists, and even a lawsuit against the company's deceptive 'carbon-negative' strategy. As the authors observe:

> Although FIJI Water has prospered in the global market and a range of positive outcomes with some positive economic benefits have been felt in the villages, a plethora of problems at a local scale have also been created. FIJI Water draws upon the constructed 'exotic' nature of Fiji to differentiate its product in a competitive global market. Yet, the places its imagery is founded upon appear to have received proportionally low benefits.[58]

---

57 C. Jones, W. E. Murray and J. Overton, 'FIJI Water, Water Everywhere: Global Brands and Democratic and Social Injustice', *Asia Pacific Viewpoint* 58: 1 (2017), pp. 112–23.
58 Ibid., 122.

Another paradox thus arises. Fiji Water's marketing strategy revolves around the idea that the water comes from a pristine and untouched place, yet the rising fortunes of the product contribute to social injustice and the environmental degradation of that same place. So, if supposedly ethically conscious companies make these blunders, what happens when Nestlé Waters – one of the biggest and most popular bottled water companies in the world – tries to seek redemption?

Corporate guilt, social redemption

Nestlé Waters – a subsidiary of the Swiss transnational food and drink giant Nestlé[59] – is the third largest bottled water company in the world, and it controls some of the most popular bottled water brands globally, including S.Pellegrino, Acqua Panna, Perrier, Nestlé Pure Life, Vittel and Contrex. In 2021, Nestlé had an operating profit in excess of 11 billion CHF.[60] Quite importantly for the sake of this discussion, over the last few decades the company has built an unenviable reputation as one of the most unethical businesses around. In 2005, for example (well before the abovementioned controversies in Osceola Township), Nestlé received the Public Eye Award[61] for 'the most blatant case of corporate irresponsibility' in relation to its 'labour conflicts in Colombia and for its aggressive marketing methods for baby food which jeopardize breastfeeding', getting twice as many votes as the second on the list (Monsanto).[62] A 2009

---

59 The world's largest food and beverage company.
60 Nestlé, *Corporate Governance Report 2021*, *Compensation Report 2021*, *Financial Statements 2021*.
61 A counter-event to the annual meeting of the WEF in Davos, during which a jury composed of NGOs and scientists elected the most irresponsible company of the year.
62 Public Eye, *'The Public Eye Awards 2005' Go To ...*, press release, 25 January 2005.

article published in the *New Internationalist* described the company:

> Nestlé is the global leader in the exploitation of water across the globe. It has 67 bottling factories and sells in more than 130 countries. In Pakistan, Nestlé, the world leader in bottled water, invented a 'blue-print factory' that could be shipped to any location in the world. It chose Pakistan for a number of reasons, one of which is that it is the only country in the region that has an unregulated groundwater sector, meaning that anyone can simply dig a hole and extract as much water as they want without paying a penny. The Pure Life water has been produced in Pakistan, Asia, Africa and South America and is marketed as 'capturing nature in its purest form'. In short, Nestlé now owns and distributes 'nature' on every continent.[63]

Nestlé's reputation did not improve. If one looks up the term 'unethical companies' on an online search engine, the name Nestlé will be displayed early on in the results. Over the last few years, Nestlé Waters has been involved in serious conflicts with local communities across the globe, most notably in Pakistan, Mexico, Nigeria, Michigan, Florida, California and in the Six Nations reserve in Canada. Many of these controversies resulted in lawsuits. In 2019, for instance, consumers from Connecticut, Maine, Massachusetts, New Hampshire, New Jersey, New York, Pennsylvania and Rhode Island sued Nestlé Waters North America, which seemingly deceived them into overpaying by labelling Poland Spring as '100% Natural Spring Water', while this was in reality ordinary groundwater.[64]

It is clear, as the company admits in a CSR report, that Nestlé Waters' 'long-term success depends on the water

---

63 S. Ekine, 'Africa: Trapped in Water Privatization', *New Internationalist*, 20 June 2011.

64 J. Stempel, 'Nestlé to Face Lawsuit Saying Poland Spring Water not from a Spring: US Judge', Reuters, 29 March 2019.

resources that supply our business operations and support the livelihoods of suppliers and consumers'.[65] Nestlé's aquifers thus need to be preserved, because once they are dry, there will be no more water to be bottled, and this is something that was also confirmed by a Nestlé decision-maker.[66] The company's 2011 vision on sustainability, which focused on water, makes no secret of this attitude:

> The ever-expanding demand for water by the world's growing, more prosperous and increasingly urbanised population, combined with the impacts of climate change policies and responses, mean that water is of increasing strategic importance for business and economic prosperity. Yet water scarcity is a reality in many parts of the world and with it, livelihoods, human health and entire ecosystems are under threat ... By 2030, demand for water is forecast to be 50% higher than today, and withdrawals could exceed natural renewal by over 60%, resulting in water scarcity for a third of the world's population ... Indeed, 70% of total global freshwater withdrawals are used by agriculture, 20% by industry, and 10% by households. Nestlé Waters uses 0.0009% of global freshwater withdrawals. It is not in the interests of our long-term business activities to mismanage the water resources we use.[67]

Nestlé Waters is indeed committed to the Alliance for Water Stewardship, a global certification for businesses and the public sector that promotes the sustainable use of water resources through the International Water Stewardship (AWS) Standard. In recent years the AWS has become increasingly popular, particularly in the private sector. And yet, a fundamental contradiction arises, particularly with regard to a bottled water company like Nestlé Waters. As the AWS website reports, 'Stewardship is about taking care

---

65 Nestlé, *Nestlé Creating Shared Value Report 2011*.
66 Interview with Nestlé Waters representative, March 2022.
67 Nestlé, *Nestlé Creating Shared Value Report 2011*.

of something that we do not own. Good water stewards recognise the need for collective responses to the complex challenges facing the water resources we all rely on.' But even though Nestlé Waters is proud of its water stewardship progress – which overall has purportedly resulted in a 2.3 million cubic meter water use reduction in its factories – and of its Caring for Water initiative, does the company really behave like one that *does not own* the water that it captures, bottles and then sells? As Peter Jones, David Hillier and Daphne Comfort remarked, much of a company's water stewardship achievements can be 'interpreted as part of a wider search for operational efficiencies and cost reductions which are driven as much by business imperatives as by any genuine commitment to the sustainability of natural ecosystems and resources'.[68] In other words, Nestlé Waters' commitments to water stewardship are entrenched within the capitalist logic of growth and technological improvements, which essentially result in what Ingolfur Blühdorn has called 'unsustainable sustainability'. Yes, I am saving water through a more efficient production process, and yes, thanks to this I can bottle and trade more water to meet a global demand for bottled water. But does that really mean that I am *saving* water? Just as the emission trading scheme set out by the Kyoto Protocol did not result in fewer emissions globally (but rather in a redistribution of emissions across countries), there is the risk that the AWS becomes yet another well-meaning initiative that companies with limited accountability use to sustain an unsustainable environmental practice.

In addition to sustainable water management, Nestlé Waters' CSR policies also focus on using fewer plastics and supporting access to safe water for communities.

---

68 P. Jones, D. Hillier and D. Comfort, 'Water Stewardship and Corporate Sustainability: A Case Study of Reputation Management in the Food and Drinks Industry', *Journal of Public Affairs* 15 (2015), p. 125.

Unsurprisingly, if we flip the reasoning upside down, these three areas correspond to the three main objections raised to the company, which is notably blamed for excessive pollution and CO_2 emissions, the depletion of aquifers and depriving communities of safe water. It might even be argued that it is guilt, above all, that drives the sustainability agenda of Nestlé Waters. When it comes to using fewer plastics, Nestlé (and the world) still has a long way to go. Globally, the company's PET water bottles are made of 88 per cent virgin plastic, and only 12 per cent comes from recycled plastic.[69] While the company's plastic packaging is getting lighter, as things currently stand, phasing out plastics is out of the question. The key, then, is to recycle more, and also in this case, there is a thin line that both unites and separates corporate from individual responsibility. Consider, for example, S.Pellegrino's[70] paradoxical – almost absurd – strategy to reduce the environmental impact of plastics. According to the company, this can be achieved through the adoption of a few simple everyday habits, including: avoid using single-use plastics (although their examples do not mention plastic bottles!); segregate waste properly; refrain from littering; and be creative when it comes to recycling, giving empty bottles a second life as vases, lamps, and pen holders (which probably explains why bottles are not considered single-use plastics). Consequently, the problem does not seem to lie in the plastic bottle itself, but rather on what consumers end up doing with it. Relatedly, the words of Mineracqua's[71] president, Ettore Fortuna, help further contextualize the matter:

---

69 Nestlé, *What Is Nestlé Doing to Tackle Plastic Packaging Waste?*

70 S.Pellegrino is a subsidiary of Nestlé Waters and one of the largest beverage companies in the world.

71 An Italian organization that protects the interests of the bottled water sector.

> We are worried about the popularity of reusable aluminium or plastic water bottles in public offices, schools, companies, hospitals and ministries ... One day someone will reflect not only about the risk of destroying a healthy sector, which employs 40,000 people, but also about the poor hygiene of these containers, and the violation of the freedom of choice of consumers: a value that cannot be cancelled.[72]

There is a clear antagonism and an ill-concealed hostility towards those who push the sector to go plastic free. It must be frustrating to see that there is an increasing political and scientific consensus about the absolute need to reduce plastic consumption, particularly if this happens when your business model is based on the availability and production of large amounts of cheap virgin plastic. On one hand, the industry wants and needs to project the image of an environmentally conscious company. On the other hand, however, the impossibility of going plastic free inevitably results in a transfer of responsibility from the company to the individual. If Nestlé cannot stop dumping its plastic on the world – Nestlé's total plastics packaging usage in 2021 was just under 1 million metric tons (around 1,000,000,000 kg) – then the consumer has to handle waste more responsibly, and of course, there is plenty of advice on how to do so coming from the company itself.[73]

So, it is not necessarily on production that we should be focusing, but rather on consumption. Consumers must become knowledgeable about plastics, since they are asked to sort and classify the various parts that they then have to separate. From caps to labels, PET/rPET to PVC, virgin to recycled plastic, a new taxonomy emerges, matched with a series of tasks for responsible and concerned consumers.

---

72 Bevitalia, *Acque Minerali, Annuario 2019–2020*, p. 10. Originally in Italian, translated by the author.
73 Nestlé, *What Is Nestlé Doing*.

And these consumers need to be introduced to the taxonomy early on. This is what 'A Scuola di Acqua' (learning about water) – an educational project funded by Levissima and S.Pellegrino – is about. Launched in 2014 with the aim of introducing participants to the importance of staying hydrated and recycling, the project has so far reached almost 400,000 children enrolled in primary schools across Italy. Among the resources available to parents and teachers, a recycling handbook provides information on how a plastic bottle is made, the benefits of PET, the meaning of various terms (such as recyclable, recycling and recycled), gives a detailed overview of the most common types of plastic (PET, HDPE, PVC, LDPE, PP and PS) and how recycling can help us save our planet. All this might lead us to think that we control plastic; rather, it is plastic that *controls us*.

Reflecting on the main achievements of the project, S.Pellegrino notes that 93 per cent of participants think that it is important to have a proactive approach to protect the environment, while 86 per cent of the children understand the importance of drinking frequently. As stated by the company, this demonstrates the achievement of the project's objectives of raising awareness about the importance of staying hydrated and recycling, 'supporting the development of critical thinking on the practice of waste sorting'.[74] Yet it is unclear if that same critical thinking should also be used to question the validity of the company's claims on tap water. According to S.Pellegrino, consumers prefer bottled water to tap water for its taste and because they think it is safer, but no evidence is provided to support this statement. In its sustainability report, S.Pellegrino compared mineral water with tap water, emphasizing how the former is supposedly purest and healthier.[75] It partly attributes this to the

---

74 S.Pellegrino, *Bilancio di sostenibilità*, 2021, p. 79.
75 Ibid., p. 55.

stainless-steel pipes used to extract it, while noting that the 'state of the public water system varies from town to town', thus implicitly saying that privately owned pipelines are in a better state than public ones. And yet, there is no mention of the 2020 European Union Drinking Water Directive, whose quality requirements are more stringent and comprehensive for water coming from the water supply network than for mineral water.[76]

Consumers need to act responsibly, as this seems to be their main role in humanity's quest to save the planet. To further engage them and develop their consciousness, Nestlé (like many other companies) launched a podcast and even a mobile app, 'Dove lo Butto' ('where do I throw it').[77] In the app, all the company's products – including its several brands of water – are listed together with the material used in their packaging (plastic is unsurprisingly ubiquitous, and its water and food preserving properties are carefully classified and described). The app is accompanied by an eponymous webzine, with the objective of raising awareness about sustainability and sharing advice on how 'together, we can reduce our environmental impact'. Nestlé – through S.Pellegrino – has also launched another webzine, 'In a Bottle', and this one is specifically about water. If one takes the time to carefully go through its articles, it seems clear that the 'global' water crisis is indeed a problem that concerns everyone, but the narrative on how to solve it has been fully absorbed, and controlled, by capital and its network of like-minded benefactors like Matt Damon.

Particularly in areas where tap water is safe to drink, bottled water is both superfluous and harmful to the environment. Where tap water is not safe to drink, bottled water can only be a temporary fix and should be distributed at

---

76 Interview with Italian civil servant, April 2022.
77 Nestlé, *Podcast Nestlé – 6 minuti di impatto positivo*, 28 December 2021.

cost price. There is no such thing as sustainable bottled water. Bottled water is underpinned by the idea that nature can be improved, which is sustained by the assumption that everything that is on (and below) the Earth's surface can be captured, owned and sold.

5
There Is No Such Thing as a Global Water Crisis

Stages of grief

In the summer of 2022, as Europe was in the midst of a severe drought, its rivers started to run dry. Their receding waters revealed artefacts that had been concealed for a long time. Among them, several 'hunger stones' – carved stones embedded into a riverbed – resurfaced, exposing inscriptions written decades and sometimes centuries before, during other periods of crisis. *Wenn du mich siehst, dann weine* ('If you see me, then weep'). Thus read one of them in the Elbe River, between the Czech and the German border. Can this be considered a moment of awakening, or even, of recognition? At the beginning of his book on climate change, *The Great Derangement*, Amitav Ghosh writes:

> Recognition is famously a passage from ignorance to knowledge. To recognize, then, is not the same as an initial introduction. Nor does recognition require an exchange of words: more often than not we recognize mutely. And to recognize is by no means to understand that which meets the eye; comprehension need play no part in a moment of recognition. The most important element of the word *recognition* thus lies in its first syllable, which harks back to something prior, an already existing awareness that makes possible the passage from ignorance to knowledge: a moment of recognition occurs when a prior awareness flashes before us, effecting an instant change in our understanding of

that which is beheld. Yet this flash cannot appear spontaneously; it cannot disclose itself except in the presence of its lost other. The knowledge that results from recognition, then, is not of the same kind as the discovery of something new: it arises rather from a renewed reckoning with a potentiality that lies within oneself.[1]

It dawns on us that the planet that we ephemerally inhabit was there before our arrival and will be there after our demise. We know that there are forces far greater than ours, and we seem to believe that just because we understand them, we can control them. We call these forces by different names, and most of these names have something do with how we come to terms with *nature*. We call the material flow of water throughout Earth the 'water cycle', and we know that our lives depend on it. We have always known this. And we have received several warnings about our increasingly unreliable – or for some, unruly – water cycle. Sometimes these warnings are quiet, and come before us rather slowly, while other times they are furiously violent. We are getting used to these warnings. But we are not losing against nature. Rather, we are *losing* a specific idea of nature, and with it, ourselves. We are *losing* a specific idea of water. And we are getting familiar with this sense of loss. We are finally starting to *recognize* it. We are beginning a grieving process. The axiom that our planet (and its human inhabitants) needs to be saved from a looming crisis is based on the notion that an originally pristine and stable nature, now disturbed, has to be restored to its original state, which needs to be preserved if ecological harmony is to be maintained. However, such a benign original state never existed. Humans *are* nature, even though we still presumptuously think that we are somehow separated from it. If this were not the case, how do we

---

[1] A. Ghosh, *The Great Derangement: Climate Change and the Unthinkable*, Chicago: University of Chicago Press, 2016, pp. 4–5.

explain that when a beaver builds a dam we call this a natural phenomenon, but when humans do, we are in front of a feat of engineering? What matters is that even though there is no way of – and no point in – knowing what the world would have been like for humans if humans were not there, we act as if we could restore an ideal type of world for humans inhabited by an ideal type of human.

Consider, for example, the notion of the Anthropocene – an unstable geological epoch in which humans and their actions are the main driver of environmental change. With it comes the acknowledgement that environmental degradation and global warming foreshadow an approaching catastrophe that will diminish, if not devastate, life for future generations. The Anthropocene brings to the surface a contradictory condition in which *anthropos* is indissolubly embedded within, and increasingly pivotal to, the environment. Moreover, this works both ways. As Rosi Braidotti notes, the notion of the human 'is not only de-stabilized by technologically mediated social relations in a globally connected world, but it is also thrown open to contradictory redefinitions of what exactly counts as human'.[2] These evolutionarily successive transformations of both our planet and its inhabitants are seen by some as an opportunity for humans to use their 'extraordinary powers' to shape a 'good' Anthropocene. In this account, human ingenuity, science and technological advancement are viewed as reasons for optimism. Yet arguments claiming that the state of humanity has apparently never been better than it is today disregard the fact that millions of people still live in extreme poverty, and that inequality is rising. Indeed, for most, environmental degradation and global warming – which are both an outcome and a reminder of humanity's inability to curb

---

2 R. Braidotti, 'Posthuman, All Too Human: Towards a New Process Ontology', *Theory, Culture and Society* 23: 7–8 (2006), p. 197.

pollution and reduce its carbon emissions – foreshadow an approaching catastrophe that will diminish, if not devastate, life for future generations. The imagination of such disasters, which Mike Davis appropriately called an 'ecology of fear', manifests in increasingly urgent warnings about the risks of 'dangerous climate change' and a wide-ranging securitization of the environment. Evoking images of the apocalypse to discuss global warming, a phenomenon essentially caused by economic growth and consumption, is evidence of the indissoluble link between capitalism and its constant state of emotional crisis. As the prevalence of apocalyptic narratives leads to their social and cultural normalization, the environment is irremediably transformed into a problem that needs to be managed through various technical fixes.

The idea of a 'global water crisis' fits into this apocalyptic narrative, and this is precisely because there is no such a thing as a global water crisis. The whole idea of a global water crisis is a highly mediatized neoliberal narrative of the Global South, created by the Global North and for the Global North, which results in a set of unfounded assumptions about water and the way it circulates across the planet. Water is a global resource as it more or less freely circulates within the Earth and its atmosphere. There certainly is one global ocean where most of the planet's water resources are stored. From there – and also from land – liquid water evaporates into water vapour, condenses to form clouds and precipitates back to earth in the form of rain and snow. So, in this sense, water is global. But at the same time, water is also heavy, expensive to transport relative to value, and 'uncooperative', both geopolitically and also as a commodity, as it is subject to market failure. But when it comes to the water crisis, water is a local and regional resource. We can talk, and we need to talk, about the freshwater crisis in Michigan, in Spain, Mexico, Mumbai, Madagascar and so on. We can also talk about the legacy of colonialism and its impacts on contemporary water inequalities in the postcolonial world. And we can talk about speculative financial markets based on water availability; the commodification and privatization of water resources; the fortunes of bottled water; underfunded water infrastructure; infrastructural inequality; uneven development; unsustainable consumption practices; urbanization and its impact on water stress; or intensive groundwater pumping for irrigation, and discuss how each of them can create a water crisis which is inevitably going to affect people differently depending on their gender, ethnicity, age, wealth or geographical location.[3] But condensing all of the above into a 'global water crisis' is tendentious and misleading. It is tendentious because it overlooks and depoliticizes the structural causes and the power inequalities behind inadequate supplies of clean water. It is misleading as it implies that the water crisis is one and the same for all, and that therefore there is one way to solve it.

And yet, even as I have cast scepticism on something as monolithic as the idea of a 'global' water crisis, this critique is challenged by climate change. This is because in its more common understanding, the 'global' water crisis spatially unfolds in a generic and distant 'developing world', and it is far from being global. But in the Anthropocene, the water crisis is paradoxically becoming global, and it is felt in places that had so far been shielded from it, as evidenced by the severe droughts that have affected Western Europe and the United States in 2022 and 2023. What is shaking the world about the water crisis is not necessarily the fact that the world is seemingly running out of water. Rather, it is the fact that the water crisis is now seriously affecting the

---

[3] See, for example, by Farhana Sultana, 'Embodied Intersectionalities of Urban Citizenship: Water, Infrastructure, and Gender in the Global South', *Annals of the American Association of Geographers* 110: 5 (2020), pp. 1407–24, and 'Water Justice: Why It Matters and how to Achieve It', *Water International* 43: 4 (2018), pp. 483–93.

wealthiest countries in the world for the first time in history. The fact that everyone is now exposed to the water (and climate) crisis in unpredictable and potentially devastating ways renders vulnerable even those who felt safe, or at least safer than the most disadvantaged – or for some, wretched – ones. So, to go back to my initial question, is this a moment of awakening? And if it is one, are we sure that by being awake things will get better?

What is evident is that as we recognize that we are losing nature, we are renegotiating our relationship with it. In her book *On Death and Dying*, Elisabeth Kübler-Ross conceptualized a series of emotions experienced by people who are grieving. She called these the five – and not necessarily successive – stages of grief: denial, anger, bargaining, depression and acceptance. Arguably, the present moment of reckoning is also a time of grieving, and the five stages of grief are suitable to analyse our current condition. We know fairly well how it feels to be at the intersection of denial and anger. Denial, on one hand, is evidenced by humanity's attitude to continue with business as usual, believing that a few technological fixes or a beer chalice – instead of a much-needed structural change – will be sufficient to lift us out of our current predicament. One might actually ask how many catastrophes we will need to finally recognize that at least we can stop worrying about a looming apocalypse, since the apocalypse has already started. Anger, on the other hand, is more multi-faceted. Environmental activists are angered and frustrated by how world leaders are not taking bold enough measures to decarbonize our economies and safeguard our ecosystems. But the corporate sector is also angry, with the bottled water industry barely able to conceal its resentment towards those who push the sector to go plastic free.

But beyond denial and anger, there is bargaining. During bargaining we might feel guilty about what went wrong in the past, and when it comes to predicting the future, we tend to assume that the worst is going to happen. This is when we start shifting responsibility and blaming others for our own sense of anxiety and insecurity. All of the above strongly resonates with the human condition in the Anthropocene, marked by a persistent obsession with looming catastrophe and unresolved disputes over who is responsible for it: is it individuals, with their consumption habits and everyday choices; multinational companies, with their devastating environmental impact; governments, who are not doing enough to avert the crisis; or perhaps our own planet Earth, Gaia, that does not meet the demands of *Homo sapiens*? And so, we are simultaneously going through anger, denial and bargaining. Then there is depression, and some – though not all – are indeed despairing, feeling helpless and hopeless about the state of the planet. Going beyond eco-anxiety, Glenn Albrecht[4] coined a neologism – solastalgia – to describe the distress or pain experienced by individuals as a result of environmental degradation, and elsewhere I have looked – together with Maja Ženko – at the link between water scarcity and depression around the desiccated Lake Urmia in Iran.[5]

We have not yet reached, however, the final stage of grief: acceptance. And we should not. With acceptance, one might feel positive about the future, more secure and relaxed. Acceptance means that we are no longer fighting against reality. But we should be fighting against reality. Not all is lost, and there is no reason why we cannot at least attempt to come out of grief while at the same time changing reality.

---

4 G. Albrecht, 'Solastalgia: A New Concept in Health and Identity', *Philosophy, Activism, Nature* 3 (2005), pp. 41–55.

5 M. Ženko and F. Menga, 'Linking Water Scarcity to Mental Health: Hydro-Social Interruptions in the Lake Urmia Basin, Iran', *Water* 11: 5 (2019), 1092.

Alternative (water)ways

It is worth interrogating the meaning of the water crisis in the Anthropocene. As I have attempted to explain, crisis is both a defining feature of our time and one of the structural foundations of capitalism. We seem nevertheless to have realized that capitalism is no longer – if it ever was – conducive to societal and planetary prosperity, and so something has to change. Throughout this book, I have endeavoured to show how capitalism manipulated three key concepts – care, sacrifice and redemption – and used them as a way of further infiltrating the water crisis. But what would happen if we were to turn around the main concerns that I have raised in this book, and prioritize the process of caring over that of profit-making? Indeed, both sacrifice and redemption revolve around care, as care triggers any activity that might lead to a sacrifice or to redemption. Care refers both to care work – for example, childcare, healthcare or domestic care work – but also to caring, 'the process of protecting someone or something and providing what that person or thing needs'.[6] Care, just like capitalism, is a process, not a thing, and it is not an end point or end in itself. Care is a dynamic lived experience.

As two recent books – *The Care Manifesto* and *The Care Crisis* – suggest, care is neglected in our current society, and this has become worse during the last forty years with the rising fortunes of neoliberal capitalism.[7] The Covid-19 pandemic has clearly exposed the effects of decades of neglect suffered by both public caring infrastructures and our economies, with the latter being understood in their

---

6 See 'care', dictionary.cambridge.org.

7 A. Chatzidakis, J. Hakim, J. Litter and C. Rottenberg, *The Care Manifesto: The Politics of Interdependence*, London: Verso, 2020; E. Dowling, *The Care Crisis: What Caused It and How Can We End It?*, London: Verso, 2022.

literal sense as 'household management'. This market logic has dismantled welfare states and democratic processes and institutions, leading to widespread austerity policies whose effects have been most felt by the poor. This also applies to water access, whereby the prioritization of profit over care has led to widening inequalities. According to the latest WHO/UNICEF report[8] on household drinking water, sanitation and hygiene, inequalities in basic service coverage between the richest and poorest increased between 2000 and 2017, and the gap between rich and poor countries is widening rather than contracting.[9]

We are surrounded by a growing army of philanthropists, concerned citizens, CEOs and celebrities, who apparently care very much (and many of them certainly mean well) about those in need of help. However, we know that access to improved water sources increases with income, and this further indicates that we value profit more than caring, even though this does not necessarily make us any happier. Are we stuck in this logic? How can we escape the perversion of care and the mechanisms that enact it? What is clear is that we cannot keep relying on individual initiatives. Throughout the book, we have seen how ongoing efforts led by individuals or private entities – be it Matt Damon, Stella Artois, Nestlé Waters or the WoWF – can at best provide a localized fix, which nevertheless continues to reproduce, normalize and legitimate the system and exploitative relations that created the need for their intervention. Many of us take for granted the fact that billionaires have to be on the frontline of the world's water crises, and that is part of the problem.

---

8 WHO and UNICEF, *Progress on Household Drinking Water, Sanitation and Hygiene 2000–2017: Special Focus on Inequalities*, New York, 2019.
9 UN-Water, *The United Nations World Water Development Report 2019: Leaving No One Behind*, Paris: UNESCO, 2019.

In Chapter 2 we looked at two water charities – Water.org and WaterAid – to examine how the act of caring can materialize on the ground. In these two cases, I have noted how both have been largely absorbed by the market and now contribute to reterritorialize water on commercial and financial circuits, leading – to varying extents – to the insulation of water from public debate and foreclosing the opportunity for active public engagement of civil society. Yet we have also seen that these charities exemplify two different approaches to caring. Water.org originated through, and places the emphasis on, individual initiative, while WaterAid is the result and expression of a collective initiative led by the UK water industry and its representatives that in turn were following the lead set by the UN Drinking Water Supply and Sanitation Decade. Both can be problematic, but the latter can at least be conducive to a broader mobilization and to the involvement – one way or another – of political and politicized communities. What is clear is that we cannot empower a self-entitled individual to act in the name of an alleged common interest.

And when that individual initiative stems from environmental activists and concerned citizens, there seems to be, as Lucas Pohl and Erik Swyngedouw suggest, a certain type of enjoyment coming from renunciation and sacrifice:

> Renouncing excessive consumption based on fossil fuels, promoting 'flygskam' (flight shame), and reducing auto-mobility as moralizing ploys, vegetarianism, recycling, and the anxiety-ridden if not depression-inducing loop of the always insufficient asceticism to make the earth and its climate whole(some) again indicate a libidinal attachment to sacrifice as a road to fullness.[10]

---

10 L. Pohl and E. Swyngedouw, 'Enjoying Climate Change: Jouissance as a Political Factor', *Political Geography* 101 (2023), 102820.

This enjoyment was evident when I discussed runathons in Chapter 3. Indeed, runathons typify some of the central dynamics triggered by the mediatized and pervasive idea of the Anthropocenic global water crisis. Usually, people decide to engage in a runathon after reading about the global water crisis on social networks. There, they are exposed to the seriousness of the problem, and they thus decide to do something about it. Runathons are a personal project, and yet their very existence is entirely dependent on the support provided by online social networks, since a runathon is driven by the narcissistic act of sharing with others an athletic achievement and a self-sacrifice (or even a pseudo-martyrdom, as in the case of Mina Guli), all the while raising money for a charitable cause and eventually finding redemption for conspicuous overconsumption. Runathons almost have a normative value: 'I am concerned about the global water crisis, I want to do something to help, and here I am, running a marathon after a few months of tough and disciplined training. I did it. So why don't you do it too?' Conversely, while those who completed a runathon have reasons to feel proud about their achievement, those who didn't might feel guilty, or even ashamed, for not doing enough. But again, we are drifting away from the process of caring and we are going back to the current, self-referential, and deadlocked set of relations between humans and water.

We need to stop feeling guilty – and consequently resentful – as this ultimately leads to further entrenchment of the current neoliberal project and the inequalities that it engenders. It is not (only) individuals who need to care or find a way out of the current mess in which we find ourselves. What we need is a collective awakening, and possibly a relational reset. There is no going back to a 'previous state of things', or to a world prior to environmental degradation or socioecological crises, since such a world never existed. We do not know how to collectively respond to the (anthropogenic) loss of nature, since we do not really know nature and the mechanisms through which *it* functions. We think that we know and understand nature and its offerings – water, wood, oil, hydrogen, oxygen and so on – but we disavow the fact that we are by all means separated from them and the processes and labour through which they reach us. Western societies are alienated from water. They only notice its existence – and the networks of pipes, taps, wastewater treatment plants, lakes, aquifers and dams through which water flows – when the resource becomes scarce, and when we feel thirsty. Humans do not feel *responsible* for nature because they do not possess the ability to respond to its loss. As Donna Haraway suggested, we need to cultivate a response-ability, we need to cultivate 'collective knowing and doing, an ecology of practices', and to do this we need to establish a new relationship with nature.[11] Nature, it is now clear, refers less to the physical world, and more to the ways in which nature emerges as a social and cultural abstraction that both stems from and feeds capitalism's material relationships and realities. Thus, we might ask, does the image of a pristine forest displayed on a water bottle label convincingly epitomize the concept of nature as a discrete node within a more-than-human assemblage? Or is this rather an image that deepens and reifies the artificial divide that separates humans from the natural environment and the entities that inhabit it? If nature is thus construed as an empty signifier for the non-human world, how does this encourage us to take responsibility for the current water crisis, if at all, other than through abstract notions of a manufactured – and overwhelmingly nostalgic – sense of loss? And perhaps most importantly, if we still insist on viewing nature as a pure landscape of mountain streams, how do we begin conceiving of a future that can exceed the human?

---

11 D. J. Haraway, *Staying with the Trouble: Making Kin in the Chthulucene*, Durham: Duke University Press, 2016, p. 34.

A possible way to cultivate a response-ability is to 'deterritorialize' ourselves, to use the words of Gilles Deleuze and Felix Guattari, or in other words, to think differently about our role as humans.[12] If we are to care for and reconnect with water, we should understand what it is that we desire, and how we can *become* water. A first step in this direction is not to think of water as something that we need, but rather as something that we need to care about. To do so, we have to learn, in schools and as a collective whole, how water circulates. We need to understand how our societies use it, and in what ways our demands can or cannot be compatible with the ways in which water works with us, not for us. Rather than focusing on water as a scarce resource, and rather than being fixated on the idea of a global water crisis, we should appreciate how abundant water is and how amazingly it manages to satisfy the needs of a society that systematically overexploits this resource. It is not water that is in a crisis. Western societies are. Once we turn the table, it is obvious that we should not repair water (or nature), and instead we should repair ourselves to become compatible with the water cycle. And as we have seen, bottled water is largely incompatible with the needs of water. Of all the subjects discussed in this book, the utter unnecessity of bottled water is possibly the most striking one. Very plainly, the fact that we continue to indulge in the large-scale trade of plastic water bottles is one of the most self-punishing and uncaring behaviours that we can legally engage in as thirsty humans of the early twenty-first century. Bottled water consumption is largely a cultural phenomenon that has greatly benefitted from the invention of the PET bottle in the 1970s. It took bottled water only a few decades to establish itself as the world's most popular beverage, and we have now

---

12 G. Deleuze and F. Guattari, *A Thousand Plateaus: Capitalism and Schizophrenia*, translated by Brian Massumi, London: Continuum, 2004.

fully normalized this habit. In the same way, we need to get used to the fact that bottled water is a luxury that we can no longer afford if clean tap water is available. For as much as imagining a world without bottled water seems difficult today, it is worth reminding ourselves how the same applied to indoor smoking before the (highly successful) European and American smoking bans that entered into force in the 2000s. Bottled water is a cultural phenomenon, and that culture has to change. Obviously, when bottled water is the only source of clean and safe water, its use should be tolerated and regulated, but we cannot allow the opposite – that for-profit bottled water establishes itself as the alternative to public water systems.

Transforming bottled water companies into public utilities can be a first step towards the construction of a renewed sense of societal responsibility. This would also be a meaningful way to own the current moment and bring forth a collective ecological consciousness whereby, as a society, we take actions that are driven by care and not by profit. Capitalism's proto-religious nature tends towards guilt and despair, and lures consumers into cause-related purchases, yet we cannot consume our way out of misery. The moment we start caring about each other for the sake of doing so is also the moment we can free ourselves from guilt and begin enjoying the present. And beyond bottled water, we need to build new forms of international solidarity and give more power and resources to strong regional and international institutions of cooperation. The entangling of the private and public sectors seems inescapable (for instance, the case of the WoWF discussed in Chapter 2), and yet, while private companies need to be held accountable for their environmental impact, they are not in charge of environmental governance. Their only entitlement in this regard is provided by the money that they make out of nature. The more money, the bigger the entitlement. But not everything that can be counted counts, and not everything that counts can be counted. It is never too late to stand up and change the discourse.

Further Reading

I have tried to write an accessible book on the water crisis, and to do so I have not delved too much into the theoretical and analytical frameworks that I have drawn upon to build my argument. In the following I provide a list of additional readings and authors on some of the key themes discussed in the book.

Nature and unsustainable sustainability

Water is nature, nature is water, and humans are nature. As Raymond Williams argued, nature is perhaps the most complex word in the English language. There are so many great books on the multiple meanings of nature that it is hard to make a list, but I will nevertheless point you to some of the authors whom I believe have made some of the most important contributions to these debates. Noel Castree's *Nature* (2005) is an accessible book that explores the shifting ways in which geographers have studied nature. Neil Smith's *Uneven Development: Nature, Capital, and the Production of Space* (1984, with a foreword by David Harvey) is not as accessible, yet it is a must-read as it foregrounds the theory of uneven development and outlines key ideas on the relationship between space, nature and capitalism. Erik Swyngedouw is one of the most original and provocative thinkers on all matters related to nature, sustainability and political ecology. If you read Spanish, there is a short open-access article of his titled '¡La naturaleza no existe!' (Nature Doesn't Exist) that demystifies a series of arguments about how to think, conceptualize and politicize nature. In another article, 'Apocalypse Forever?', published in the journal *Theory, Culture and Society*, Swyngedouw draws heavily on Alain Badiou to advance a critique of environmental catastrophism and certain forms of environmentalism. There is also a more recent article (co-authored with Lucas Pohl) published in the journal *Political Geography* titled 'Enjoying Climate Change: Jouissance as a Political Factor', in which the authors provocatively discuss how climate change can be interpreted not only as a planetary threat but also as something that is enjoyed in one way or another by both big polluters and environmental activists.

Water

If, after reading this book, you want to know more about water, I would recommend Karen Bakker's *Privatizing Water: Governance Failure and the World's Urban Water Crisis* (2010), which provides an excellent introduction to the pitfalls of the neoliberalization of water resources across the globe. There are three books that I read many years ago that greatly influenced my understanding of the inseparable link between water and political power: Mark Reisner's *Cadillac Desert* (1986); Donald Worster's *Rivers of Empire: Water, Aridity, and the Growth of the American West* (1992); and Timothy Mitchell's *Rule of Experts: Egypt, Techno-Politics, Modernity* (2002).

Capitalism and religion

Walter Benjamin's 'Capitalism as Religion' has informed a number of important works that explored the contradictory and parasitical relationship between capitalism and religion. Michael Löwy's essay 'Capitalism as Religion: Walter Benjamin and Max Weber', published in the journal *Historical Materialism* in 2009, is a good entry point into some of the issues discussed in this book. The work of Luigino Bruni, an Italian political economist, is a must-read for anyone interested in how modern capitalism appropriates and transforms the Christian spirit into the spirit of capitalism. His books and articles – generally available in both Italian and English – bring together approaches from political theology and political economy to convincingly (and engagingly) explain how this parasitic relationship unfolds.

Sacrifice

The deeply spiritual concept of sacrifice is key to understanding the profound – and often twisted – meaning of a donation. There are several interrelated French authors who have explored these themes. Georges Bataille's *La part maudite* (1949) is one of the first books that developed a theory of consumption through a reflection on how societies deal with the 'excessive' or 'exceeding' part of any economy (crucial to this theory is the phenomenon of the *potlatch*). The work of Bataille has strongly influenced René Girard's *La violence et le sacré* (1972) and has also informed Gilles Deleuze's and Félix Guattari's *Anti-Oedipus*.

Care

There is a growing and heterogenous range of books that do not take 'care' for granted, discussing its meaning and the value of caring in the current times, and this is symptomatic of what Emma Dowling called *The Care Crisis* (2022). *The Care Manifesto*, written by the Care Collective and published in 2020, uses care as an analytical lens to discuss the roots of the current socioecological crisis, and advance a vision for a truly caring world. Likewise, María Puig de la Bellacasa's *Matters of Care: Speculative Ethics in More Than Human Worlds* (2017), engages with scholarship on the non-human/more-than-human to also challenge and interrogate the shifting ethico-political meanings of care and its implications in various contexts.

Philanthropy and philantrocapitalism

There are several excellent books that look into the drives and unintended consequences of philanthropy, and that critique the workings of the growing army of do-gooders in international development. There are two books on the celebrated (and celebrity) economist Jeffrey Sachs that are recommended to all those willing to learn more about the shortcomings of one-size-fits all approaches, that might work on paper but not in the real world. These are Nina Munk's *The Idealist: Jeffrey Sachs and the Quest to End Poverty* (2014) and Japhy Wilson's *Jeffrey Sachs: The Strange Case of Dr. Shock and Mr. Aid* (2014). Linsey McGoey's book on Bill Gates, *No Such Thing as a Free Gift: The Gates Foundation and the Price of Philanthropy* (2015), puts the current golden age of philanthropy under scrutiny showing its complexities and limitations, and this is also brilliantly done by Nicole Aschoff in *The New Prophets of Capital* (2015).

Acknowledgements

This book is the modest result of more than five years of thinking and writing. Five years also means meeting hundreds of people, reading thousands of articles and books, and sending and receiving probably more than 30,000 emails. Writing is probably the recreational activity that gives me the most pleasure. Whether or not I am good at it is another matter. But I like to do it, partly because the written word, especially on paper, is something that lasts. And that gives a partial sense of fulfilment to me, a consciously chronophobic person. All this is to say that there are some people I would like to thank because, if you are holding this book in your hands, the credit is also theirs. Unfortunately, there is no way I can remember and acknowledge all the people who have helped to shape and refine my argument. But a few stand out.

On the editorial side, I would like to thank Nicole Aschoff for believing in the project and supporting me in its early stages. She was my point of entry into Verso, and I could not think of a better person, not least because she is the author of a wonderful book on the many contradictions of capitalism (*The New Prophets of Capital*). At Verso, I would also like to thank Asher Dupuy-Spencer and Kelly Burdick, who saw the manuscript through to publication and made it a better read overall. I would also like to thank my agent, Fiammetta Biancatelli, for her support and advice on navigating the publishing world. One particular academic, Mike Goodman, my former mentor at Reading and now a friend, played a crucial role in the genesis of this idea, and I am grateful for his support.

Others, and I mention them in no particular order, have also left a mark in one way or another, and I am grateful to them: Tom Grisaffi, Naho Mirumachi, Maria Rusca, Lyla Mehta, Erik Swyngedouw, Francesca Greco, Rossella Alba, Alberto Vanolo, Filippo Celata, Giovanni Sistu, Maja Zenko, Rico Isaacs, Annalisa Addis, Lisa Ann Richey, Luisa Cortesi, Alessandro Uras, Emanuele Leonardi, Danny Waite, Federico Cugurullo, Mette Riise and Giorgio Osti. There are many more that I am forgetting.

Then there are my colleagues at the University of Bergamo (and formerly at the University of Reading, as this book does indeed go back a long way), who have given me the freedom to concentrate on this project.

Family, friends: you, as above, know who you are. Years ago, when writing the acknowledgements in another book, I apologized for being a bit absent and named you all. I do not do so here, but I hope I have been a little more present in your lives.

Finally, the sea. That's where I started, and that's where I hope to end.

Index

access, 2, 21, 46
Affinity Water 85
Africa
  and capitalism 36
  modernization attempts 34–7
  water infrastructure 34–7
Agamben, Giorgio 52, 101
Agenda 2030 49
agriculture 19–20, 165
aid
  celebritization of 96
  fetishization of 105–6
  neoliberalization of 93
Algeria 35
Alliance for Water Stewardship 153–4
Alternative World Water Forum 32
Amazon 73, 74
Annan, Kofi 43–4
Anthropocene, the xi, 24, 30, 71, 113–14, 141, 163, 165–6, 167
apocalyptic narrative 161–7
AquaFed, International Federation of Private Water Operators 49
Augustine, St 71–2
austerity policies 169
Australia 3n1, 37–8
availability crisis 21
awareness raising 3–5, 117–18

Bakker, Karen 16, 47, 85–6, 130
Ban Ki-moon 43–4

Band Aid 96
Bangladesh 144
Bank of America 79
Barbier, Ed 57–8
Bataille, Georges 106, 107
Baudrillard, Jean xii, 53
Belgium 144
Belu Water 116
Benjamin, Walter, 'Capitalism as Religion' 51–3
Bezos, Jeff 73–4
Bezos Earth Fund 73, 74
biodiversity loss 8n5
biopolitics 60
Blühdorn, Ingolfur 154
Boer, Roland 55, 55–6
Bono 68, 68–9, 71, 97, 99, 101, 113
bottled water 80, 130, 165, 166
  clean presentation 139
  consumption 5, 125–30, 130, 145
  contamination 139, 143–4
  Corporate Social Redemption 148
  Corporate Social Responsibility 139, 145–51
  cost 131–2
  environmental impact 174
  as essential public service 134–5
  ethically branded 149–51
  global trade patterns 144–5
  growth 130–1

INDEX

bottled water (*continued*)
   as non-essential product 145
   perception of 131
   plastic pollution 140–5
   plastic-free transition 143
   sales 129
   success 126
   sustainability 149–51, 153, 158–9
   transformation of industry 173–5
   and water scarcity 132–9
Boutros-Ghali, Boutros 43–4
Brabeck-Letmathe, Peter 135–6
Braidotti, Rosi 163
Brand Aid 101
Branson, Richard 113
Brazil 20–1
brewing and the brewing industry 12–15
British National Water Council 41
Bruni, Luigino 107
Bush, George W. 68

canals 133
Cape Town x, 2, 10–12
capital, and water crises 57–8
capitalism xii, 9
   absorption of the water crisis 50–1
   and Africa 36
   and charity 89–90, 122
   and crisis 2, 122, 164, 168
   and global water crisis 89–90
   great achievement 53
   logics of growth 1–2
   manipulation of sacrifice 116–17, 123
   marketization logics 46–7
   Marx's view of 53
   as religion 51–6, 174
   and religion 106–7, 107–14
   self-preservation 148

   spirit of 51–6
   symbolic sacrifices 106
   taboos 106–7
capture 130–1
care and caring 23, 59–61, 66, 92–3, 99, 102, 168–9
cause-related marketing 119–20
celebrities and celebrity-led campaigns 23–4, 56, 67–71, 74–5, 77–9, 81–2, 93, 95–102, 113–14, 116
charities
   analytical framework 22
   expenditure 63–4
   growth in numbers 63–4
   political economies 97
   see also global water charities
Charities Aid Foundation 65
charity 1
   and capitalism 89–90, 122
charity donations
   economic capital 114–22, *115*
   as a fetish 56, 105–6
   individual action 108–14
Charles III, King 83
Chile 19
China 3n1, 122, 144, 150
cholera 20
civil society, foreclosing 170
CK Hutchison Holdings Limited 88
climate change 7–8, 16, 165
Climate Clock 30
climate crisis xi, 24, 30–1, 74, 83–4, 163–4
climate modelling 40
Clinton, Bill 68, 74–5
Clooney, George 97
Club of Rome 38–41
CME Group 120–1
CO_2 emissions 31, 74, 145, 150, 155, 164
Cold War, end of 40
Colgate, Super Bowl commercial 3–4

INDEX

collective responsibility, and individual action 107–14
colonial thinking 93
colonialism 12, 164
Colorado River 13
commercialization 47, 50–1
commodification 8, 45, 50, 80, 81, 92, 122–3, 130–1, 138–9, 165
commodity fetishism 54–6
commodity frontier, expansion of 2, 138–9
common interest 170
commons 45–6
  water as 135–6
Commonwealth Department of Health, Australia 37–8
compassion, commodification of 114–22, 115
conflict, and water scarcity 43–4
Conrad N. Hilton Foundation 79
Constellation Brands 10, 12–15
consumer choice 99
consumption 4–5, 54, 100, 164, 165, 167
  awareness raising 117–18
  bottled water 5, 125–30, 130, 145
  industrial 19–20
  redistributive effects of 6
consumption of consumption xii
contamination
  bottled water 139, 143–4
  drinking water 30
  faeces 20
  microplastics 143–4
COP26 147
COP27 146–7
Copernicus Emergency Management Service 103
Corporate Europe Observer (journal) 137–8
corporate giving 87
corporate philanthropy 93

Corporate Social Redemption 148–51
Corporate Social Responsibility 90, 101, 139, 145–51, 151–9
corporate sponsorship 100
corruption 35
Courage and Civility Award 73
Covid-19 pandemic 6, 61, 73, 168
credit 52
crisis, and capitalism 2, 122, 164, 168
crisis management, profit-making 37
cultural capitalism 126
cultural geographies 149

Dakar Declaration 29
Damon, Matt 1, 22, 67–71, 74–5, 75–6, 77–8, 81–2, 113–14, 115, 117, 169
Danone 128
Davis, Mike 21, 125, 164
de Brosses, Charles 54
decarbonization 166
Declaration of the International Drinking Water Supply and Sanitation Decade 41
Deleuze, Gilles 173
Demeritt, David 16–17
dépense 106
deregulation 65, 138
desalination ix–x, 10, 11
developing world, the, structural problems 105
developmentalism 82
dispossession 57
distribution 42
doctrine of the two cities 71–2
doctrine of the two kingdoms 72
Draghi, Mario 6
drinking water contamination 29–30
Duarte, Jesus Galaz 14–15

INDEX

Dublin Conference, 1992 42–3
Dublin principles 42–3, 47

eco-anxiety 167
ecological degradation 6
ecology of fear 164
economic capital, charity donations 114–22, *115*
economic choice, biopolitics of 60
economic equity 48
economic good 47
economic growth 164
Economist (journal) 61
Elbe River 161
empathy 97, 99
enclosure 130–1
energy crisis 6–7
energy prices 6–7
Enterococcus faecalis 139
environmental change 163
environmental crisis xi, 2, 24, 40
environmental degradation 8n5, 83–4, 151
environmental forecasting 40
environmentalism 82
Escherichia coli 139
Ethos Water 149
Europe, drought, 2022 103–44, 161
European Union 7, 86
European Union Drinking Water Directive, 2020 158
everyday moralizations 59
everydrop-counts.org 3–4
Evian 127–8, 150
Évian-les-Bains 127
extreme weather events xi

faeces contamination 20
Fauchon, Loïc 27, 28
Ferragni, Chiara 128
fetishism 54–6
fetishization 105–6, 119

Fidelity® Water Sustainability Fund (FLOWX) 121–2
FIDES 36
Fiji Water 150–1
financial recession, 2007–8 2
Financial Times (newspaper) 87
Finish 4
finite resources 18
Flint water crisis 2, 126, 131
floods xi
Ford Foundation 108
formula feed 127
Fortune 111
fossil fuel industry 146–7
France 128
  bottled water exports 144
  colonial empire in Africa 34–7
  FIDES 36
  Oudin-Santini law of 2005 33–4
freshwater resources, distribution 20–1
fundraising 66, 107
  celebrity-fronted 100
  individual action 110–14
futures market 120–1, 165

Gates, Bill x, 62, 69–70, 70–1
Germany xi, 7, 126, 144
Ghosh, Amitav 161–2
gifts 106–7, 116–17, 119–20
Girard, René 107, 119
glacial melting xi, 2
glass bottles 143–4
Global North 39, 59, 164
Global South 50, 50–1, 105, 164
global trade patterns, bottled water 144–5
global warming xi, 83, 163, 164
global water charities 122–3, 170
  commitment 65
  critique of 56, 63–7
  dependence on international donors 64
  emergence of 59–67

INDEX

expenditure 64
funding 66
growth in numbers 63–4
influence 63
market logics 90
origins 84
relationship with capitalism 90
research on 65–6
variety 65–6
see also WaterAid; Water.org
global water crisis 1–22, 37, 92–3
  apocalyptic narrative 164–5
  awareness raising 3–5
  and capitalism 89–90
  and climate change 7–8
  decontextualized 29
  deployment 37
  first recorded use of the term 37–8
  growth and circulation of concept 37–45
  inevitability 121
  interconnections 15–16
  mediatized 171
  narrative 33–4
  profiting from 58
  range of issues 44
  social and political dynamics 18–21
  social construction of 16–17
  solution 8–9
  spread of exposure 165–6
global water governance 27–31, 42, 83–4
Global Water Partnership (GWP) 28, 43
Global Water Partnership South America 112
Global Witness 146–7
glocalization 132–9
Go Water<Less campaign 117–18
good intentions 23

Good Samaritans 67–75, 85, 113
governance 9, 24, 27–31, 42
governance beyond-the-state 47–8
Great Pacific Garbage Patch 141
green neoliberalism 23
greenhouse gas emissions 140, 145
greenwashing 146–8, 148–9
grief, stages of 166–7
growth, logics of 1–2
Guattari, Felix 173
guilt 171–2, 174
Guli, Mina 110–13, 171
Guterres, António 7, 147

H&M 116
H&M Foundation 116
H2O Africa Foundation 69, 76, 81–2
Harmon, Judson 13
Harmon Doctrine, the 14
Harpic 118
Health (journal) 37–8
Heineken Foundation 116
household management 169
HSBC 116
human needs 42
human rights 136
humanitarian imaginary, media performances of 96–7
humanitarianism
  celebrification of 96–100
  visualities of 102–3
humanity, state of 163–4
hunger stones 161
hydration 157
hydrosocial contract 48

iCare capitalism 60
ideological mechanisms 93
ideological perspectives, NGOs 64–5
images, as social actants 102–3
imperialism 73

187

INDEX

income 169
India 122, 131
individual action 3–5
  and collective responsibility 107–14
  role of 104
individual responsibility 104
industrial water consumption 18–19
inequality 30, 73, 91, 108, 163, 165, 169
infrastructure
  inequality 165
  resilience 8
  underfunding 165
innovation 28
intellectual Atlanticism 39
Intergovernmental Panel on Climate Change (IPCC) 7
International Decade for Action 'Water for Sustainable Development' 49
international donors, dependence on 64
International Drinking Water Supply and Sanitation Decade 83
International Federation of Private Water Operators, AquaFed 49
International Monetary Fund (IMF) 48, 50
International Union for the Conservation of Nature (IUCN) 27
International Water Association (IWA) 27
International Water Resources Association (IWRA) 27, 41–2
International Water Stewardship (AWS) Standard 153–4
investment, water services 137
investment opportunities 120–2, 123
invisible water 111

irrigation 19–20, 165
Israel, drought, 2018 11
Israeli Water Authority 11
Italy xi, 4–5, 5, 6–7, 20, 128–30
  bottled water consumption 126
  bottled water exports 144
  bottled water sales 129
  drought, 2022 103–4

Japan 144
jeans 117–18
Johnson, Boris 142
Jolie, Angelina 97

Kidman, Nicole 150
King, Alexander 39
Kübler-Ross, Elisabeth 166–7
Kyoto Protocol 154

Lacan, Jacques 53
Laos 144
Latouche, Serge 53
leaks 5
Levi's® 117–18
liberal philanthropy 108
life expectancy 128
Live Aid concerts 96
localized solutions 105
long-term targets, effectiveness 83–4
Löwy, Michael 53
Luther, Martin 72
Lysol 118

Madonna 97
Malawi 90
market environmentalism 47
market failure 164
market logics 90
market-driven solutions 74–5
marketization logics 46–7
Marseille, WoWF 30, 31–4
Marshall, Sir Robert 85
Marx, Karl 53–5

INDEX

Médecins Sans Frontières 96
Medicine, Conflict and Survival (journal) 137
mega dams 133
messages xii
Mexicali, Constellation Brands deal 10, 12–15
Mexicali Resiste 13, 14–15
Mexicali Valley aquifer 13
Mexico xi, 5, 125–6, 131–2, 132–3
  Constellation Brands deal 10, 12–15
microfinance 77–9
microplastics 87, 141–2, 143–4
Millennium Development Goals' (MDGs) 49
Millennium Villages Projects 70
mineral water, therapeutic effects 126
modernity 51–2, 53, 56
Momoa, Jason 95
money 52, 55, 58
Monsanto 151
moral authority, celebrities 96–7, 100
moralistic education 97
More than a Toilet campaign 118
Morocco 35
Mozambique 90
Musk, Elon ix

narrative, global water crisis 33–4
Nasdaq Veles California Water Index futures (NQH2O) 120–1
National Geographic (journal), 'make a difference' campaign 4–5
national security 43–4
National Water Council 82, 85
nature
  alienation from 172
  humans and 162–4
  limits of 18–19
  recognition of loss of 161–7, 172
  relationship with 25, 172
  responsibility for 172
Nature Sustainability (journal) 18
neo-colonialism 35–6, 98
neoliberalism 42, 71–4, 99
  care and caring under 60–1
neoliberalization 45, 63, 66, 93, 102
neo-Malthusian theories 18–19, 38, 38–9
Nestlé 104
Nestlé Waters 129, 169
  Alliance for Water Stewardship commitment 153–4
  Caring for Water initiative 154
  controversies 152
  Corporate Social Responsibility 139, 151–9
  Dove lo Butto app 158
  objections against 155
  operating profit 151
  plastics packaging usage 156
  plastics use policy 154–7
  Public Eye Award 151
  reputation 151–9
  sins 24
  and sustainability 22, 56, 153, 158
  and water scarcity 134–9
  water stewardship 153–4
net-zero commitments 147
New Internationalist (journal) 152
NGOs 32, 47, 60, 63, 64–5, 96, 100, 136
Niagara Bottling 79–81, 132–3, 143
Nigeria 139
North West Water Authority 83
North–South relations 23

Northumbrian Water 88
NQH2O 120–1

Obama, Barack 150
Organization for Economic Co-operation and Development (OECD) 39, 140–1
  Assistance Committee 65
Osceola Township 134–5, 138–9
Oudin-Santini law of 2005 (France) 33–4
overcrowding 21
Oxfam 84, 100

Pakistan 139
Parker, Sarah Jessica 116
Peccei, Aurelio 39, 40–1
philanthrocapitalism 24, 61–2, 96, 99
philanthropic capitalism 71–4
philanthropy
  corporate 93
  liberal 108
  philanthrocapitalism 24, 61–2, 96, 99
  philanthropic capitalism 71–4
  popular philanthrocapitalism 62
  symbolic 107
place, sense of 130
plastic bottles 80, 81, 125, 126, 128, 140–5, 155–7, 173
plastic consumption 140
plastic pollution 81, 140–5
plastics, recycling 140, 141, 142, 143, 155–7
plastics production 141
plumbing poverty 44
Po River 103
policymakers 99
pollution 164
polyethylene terephthalate (PET) 81, 128, 140–5, 143, 155–7, 157, 173

popular philanthrocapitalism 62
population 8, 18–21
Portsmouth Water 85
Portugal 126
post-humanitarianism 97
potlatch 106, 107–8
#PourItForward campaign 116–17
poverty 163
power relations 89, 91, 123, 165
prices and pricing 48, 86, 92, 131–2
private sector participation 47
privatization 14, 23, 43, 47, 48–9, 85, 88–9, 130, 131, 138, 165
privatization decade, the 50
privilege 108
Product (RED) campaign 101
production-for-profit system, exploitative nature of 148
profit and profit-making 87, 137, 169
projections, effectiveness 83–4
pseudo-martyrdom 110–14
Pseudomonas aeruginosa 139
Public Eye Awards 151
public water, mistrust in 131
public-private partnerships 66

racism 2
rainfall xi, 18
rational planning, diffusion of 42
rationing x, 104
Reckitt 118
Red Cross 100
Reformation, the 72
relational reset 171–2
religion, and capitalism 106–7, 107–14
re-municipalization 46
renewable energy 7
reservoir effect, the 10
reservoirs, levels x
response-ability, cultivating 172–3

INDEX

Rio Grande 14
Rodney, Walter 35–6
ruins, reappearance x
runathons 110–13, 171
Running the Sahara (documentary) 69
Russia, invasion of Ukraine 6

S.Pellegrino 104, 128–30, 155–8
Sachs, Jeffrey 68, 69
sacrifice 99, 106, 107–14, 168
   capitalist manipulation of 116–17, 123
   enjoyment of 170–1
   political economies of 101–2, 119–20
Salini, Pietro 103
Sall, Macky 27
San Pellegrino Terme 128–30
sanitation 2, 17, 20, 21, 29, 33–4, 35, 41, 46, 49, , 74–5, 75–9, 81, 83–5, 87, 88, 90, 99, 100, 113–14, 118, 169–70
Sanpellegrino S.p.A 128–30
Sardinia x, xi
Sarkozy, Nicolas 33–4
Save Darfur 96
Save The Children 84
schistosomiasis 20
science 163
Seales, Paul 68
self-sacrifice 108–10, 171
Senegal 35–6, 144
service coverage 30, 169
SES Water 85
sewage 87
shopping choices 101
sign value 53–4
single-use plastic packaging 140–1, 155
slums 21
smart investment 77–8
smoking bans 174
social actants, images as 102–3

social commons 46
social injustice 151
social media 110, 171
social metabolism 74
social networks 109
social water crisis 102–7
Société anonyme des eaux minérales de Cachat 127
Société des Eaux de Marseille 27
socio-ecological crisis 59
solastalgia 167
South Africa x, 2, 10–12
South East Water 85
southern Europe, drought, 2022 103–5
spa towns 127
spectacularization 31, 95–6, 114
Starr, Ringo 23
state, the, hollowing out of 47, 63
state failure 46–7
state–society relations 45
Stella Artois 1–2, 15, 116–17, 169
stewardship 153–4
Stockholm International Water Institute (SIWI) 28, 43
#strikewithme 113
structural dynamics 123
structural problems, developing world 105
structural water crisis 104–5
Suez 34–6
Suez Lyonnaise des Eaux 27
Super Bowl 1–4, 117
super-rich, the, growth of 61
supply and demand cycle 10–11
sustainability
   bottled water 149–51, 153, 158–9
   Nestlé Waters and 22, 56, 153, 158
   production of unsustainability 15
unsustainable 154

INDEX

Sustainable Development Goals 17
symbolic philanthropy 107

taboos 106–7
tap water 1, 5, 86, 126, 128, 131, 139, 144, 157–9, 174
technology 28, 163
Thailand 125
Thames Water 88, 88–9
Thatcher, Margaret 85–6
thermotherapy 127
Thirst Foundation 110–14
Thirst Foundation Six 112
Thirsty Third World (TTW) conference 41, 82–4
Thunberg, Greta 97
toilet strike 113–14
toilets 118
transnational markets 89–90
transport 16
trust, loss of 131
Tucci, Stanley 129
Turkmenistan 20

Ukraine, Russian invasion of 6
UN Climate Change Conference of the Parties 146–7
UN Conference on Environment and Development (UNCED) 42–3
UN Development Program (UNDP) 27
UN Drinking Water Supply and Sanitation Decade 85, 170
UN Educational Scientific and Cultural Organization (UNESCO) 27
UN Environment Programme's (UNEP), Advocate for Life Below Water 95
UN General Assembly 41
UN International Decade for Action on 'Water for Life' 49
UN International Year of Freshwater 49

UN Ocean Conference 95
UN Water Conference, 1977 38, 41, 82–3
UN Water Conference, 2023 112
uneven development 165
United Kingdom
    charities 63–4
    drought, 2018 x–xi
    investment cut 87
    microplastic pollution 87
    water companies ownership 88–9
    water prices 86
    water quality 86
    water supply privatization 85–7, 88–9
United Nations 17, 21, 42, 125
    Agenda 2030 49
    funding streams 100
    initiatives and campaigns 49
United States of America
    bottled water consumption 125
    bottled water imports 144
    Flint water crisis 2, 126, 131
    Osceola Township 134–5, 138–9
    plastic bottles 144
    plumbing poverty 44
    racism 2
    water pumping rights 132–3, 134–5
    water withdrawals per capita 3n1
unsustainability, production of 15
unsustainable sustainability 154
UN-Water 20
UN-Water High Level Political Forum 112
urbanization 21, 125, 126, 165
urgency, sense of 29–30, 57
USSR 16–17, 133n20
Uzbekistan 20

INDEX

Veolia 34
Veolia Water 27
virtual water 111
visualities, of humanitarianism 102–3
volunteering 107

warnings 162
　effectiveness 83–4
waste exports 144
waste mis-mangement 144
wastewater 20, 87
water
　abundance 173
　amount 18
　centrality of 68–9
　characteristics 16
　finite 18
　loss of idea of 162
　reconnection with 173
　social and political dynamics 18–21
water abstraction 19n31
Water Aid 41
water capture 43
water charities 22, 22–3
water congresses 28–9
water coverage, extension 46
water crisis x–xi
water cycle 162, 164
water discourses 8–9
water disputes 8
water footprint 4–6
water governance, neoliberalization 66
Water International (journal) 42
water management 9
　commercialization 50–1
　as commons 45–6, 135–6
　global 45–51
　governance beyond-the-state 47–8
　political dimension 57–8
　privatization 48–9
　state failure farming 46–7
　technocratic approaches 57
water meters 35
water network 22
water pumping rights 132–3, 134–5
water quality 2, 5, 86
water resources, appropriation of 57
water scarcity 132–9, 138
　absolute 3
　and conflict 43–4
　sense of urgency 2–3
water sector, growth 122
water security 8, 27–8
water services
　delivery of 45
　investment 137
water stress 18, 21
water treatment 9
Water Users Associations 92
water wars 2, 8, 34, 43–4
water withdrawals, global, 1950s 19
WaterAid 22, 23, 56, 66, 82–92, 170
　capitalist logics 91–2
　cognitive legitimacy 90
　foundation 66, 84–5
　funding 66
　'Give it up for Taps and Toilets' campaign 108–10
　income 87
　income sources 114–16, *115*
　influence 88
　membership 88
　mission 88
　multinational corporation partnerships 115–16, 116–18
　neoliberal logics 66
　reliance on celebrity campaigns 100
　reliance on donors 90
　reports 90–1

WaterAid (*continued*)
  strategy 85, 90
  tension within 90–1
WaterAid Malawi 92
WaterAid Nigeria 91
WaterCredit Initiative (WCI) 76–7
WaterCredits 117–18
WaterEquity Global Access Fund 79–81
Water.org 1–2, 15, 21–2, 23, 56, 66, 67, 75–82, 85, 100, 132, 170
  aim 75, 76
  foundation 66, 74–5, 75–6
  funding 66
  income sources 115, *115*, 116–18
  market logics 67
  market-driven solutions 74–5, 76–9
  microloans 77–8
  mission 79
  revenue and expenses 77
  shop to support platform 119–20
  WaterCredit Initiative (WCI) 76–7
WaterPartners International 76
water-related diseases 18–19, 20

wealthiest 1 per cent, CO_2 emissions 74
Weber, Max 51, 53, 72
Webuild 103
Wessex Water 88
White, Baroness Eirene 84
White, Gary 67, 68, 74–5, 75–6, 81–2, 115
white saviour trope 61, 97–8
WHO 21
World Bank 23, 27, 42, 48, 50
World Water Congresses 42
World Water Council (WWC) 27, 28, 43
World Water Day 116–17, 117–18
World Water Forum (WoWF) 27–9, 31–7, 43, 169, 174
World Water Forum (WoWF), 2000 136–8
World Water Forum (WoWF), 2024 ix
'World Water Vision: Making Water Everybody's Business' 137
The Worth of Water (White and Damon) 67, 68, 74–5, 81–2

Young Global Leaders 111

Zambia 67, 68–9